Among African Apes

Among African Apes

Stories and Photos from the Field

Edited by

Martha M. Robbins
and Christophe Boesch

UNIVERSITY OF CALIFORNIA PRESS
Berkeley Los Angeles London

University of California Press, one of the most distinguished university presses in the United States, enriches lives around the world by advancing scholarship in the humanities, social sciences, and natural sciences. Its activities are supported by the UC Press Foundation and by philanthropic contributions from individuals and institutions. For more information, visit www.ucpress.edu.

University of California Press
Berkeley and Los Angeles, California

University of California Press, Ltd.
London, England

Library of Congress Cataloging-in-Publication Data

Among African apes : stories and photos from the field / edited by Martha M. Robbins and Christophe Boesch.
p. cm.
Includes bibliographical references and index.
ISBN 978-0-520-26710-7 (cloth : alk. paper)
1. Apes—Africa—Anecdotes. 2. Apes—Africa—Pictorial works. 3. Apes—Research—Africa—Anecdotes. I. Robbins, Martha M., 1967– II. Boesch, Christophe.
QL737.P96A464 2011
599.88096—dc22 2010033131

Manufactured in the United States of America

20 19 18 17 16 15 14 13 12 11
10 9 8 7 6 5 4 3 2 1

In keeping with a commitment to support environmentally responsible and sustainable printing practices, UC Press has printed this book on Rolland Enviro100, a 100% post-consumer fiber paper that is FSC certified, deinked, processed chlorine-free, and manufactured with renewable biogas energy. It is acid-free and EcoLogo certified.

CONTENTS

ILLUSTRATIONS

PLATES *(following page 16)*

BOXES

ACKNOWLEDGMENTS

The details of the chapters in this book are the result of an accumulation of decades of work at all the different field sites. Therefore, we first want to allow all the contributors to acknowledge the contributions of government agencies, donors, assistants, colleagues, and friends who helped make the work possible.

Christophe Boesch thanks the Ministère de la recherche scientifique as well as the Ministère des Eaux et Forêts, the Direction de la Protection de la Nature, and the Office Ivoirien des Parcs et Réserves for permission to work in Taï National Park and for their constant and friendly support during the thirty years of the project, as well as the direction of the Taï National Park and the Swiss Centre for Scientific Research for supporting the Taï chimpanzee project. Christophe Boesch thanks many foundations and particularly the Swiss National Foundation and the Max Planck Society for constant financial support. The Taï chimpanzee project would neither have existed nor developed without the dedicated contributions of numerous students and field assistants during its thirty years of existence, and they are all thanked particularly warmly.

Gottfried Hohmann and Barbara Fruth thank the Ministère de l'Education nationale, Secrétariat général de la recherche scientifique et technologie, and the Institut congolais pour la conservation de la nature for granting permission to conduct research at Lomako and Salonga. The financial support from the Max Planck Society, the Federal Ministry of Education and Research of Germany, the Leakey Foundation, National Geographic Society, Volkswagen Foundation, and private donors is gratefully acknowledged. Sincere thanks go to the people from the villages of N'dele and Lompole, respectively, for their support and friendship, and to all those who have helped us do field research on bonobos in the Congo basin over the past twenty years.

Thomas Breuer and the Mbeli Bai Study are grateful to a variety of dedicated scientists who helped to initiate and expand the study, notably (ordered by date): Steve Blake and Mike Fay, Claudia Olejniczak, Yako Valentine, Richard Parnell, Dave Morgan, Tina Goody, Emma Stokes, Angela Nowell, Mireille Breuer-Ndoundou Hockemba, Lyndsay Gale, Vicki Fishlock, Ella Emeline Bamona, Joel Glick, Franck Barrell Mavinga, Julia Jenkins, Kelly Greenway, and Doreen Schulz. Thanks go to the Congolese Ministry of Sustainable Development, Forest Economy and Environment and the Department of Scientific Research and Technology for permission to work in the Nouabalé-Ndoki National Park. Without the support of the Wildlife Conservation Society, especially all the personnel of the Nouabalé-Ndoki Project, it would not have been possible to concentrate fully on the activities specific to the project. The Mbeli Bai Study is particularly grateful to the following donors: the Brevard Zoo, Columbus Zoo and Aquarium, Cincinnati Zoo and Botanical Garden, Cleveland Metropark Zoo, Disney Worldwide Conservation Fund, Houston Zoo, Margot Marsh Biodiversity Fund, National Geographic Society, Sea World & Busch Gardens Conservation Fund, Toronto Zoo, United States Fish and Wildlife Service, Wildlife Conservation Society, the Woodland Park Zoo, and several other zoos.

Cleve Hicks would like to thank the Ministère de l'Environnement of the Democratic Republic of the Congo for granting him permission to work in the country, and Chief Zelesi Yakisi for acceptance into his collectivity. Thanks go to Steph Menken, Jan Sevink, Hans Breeuwer, and Peter Roessingh of the Institute for Biodiversity and Ecosystem Dynamics at the University of Amsterdam for their continuous support of the research. Thanks as well to Christophe Boesch, Martha Robbins, Gottfried Hohmann, Hjalmar Kühl, Geneviève Campbell, Sandra Tranquilli, and the great team of researchers at the Max Planck Institute for Evolutionary Anthropology for providing invaluable assistance and feedback. The Wasmoeth Wildlife Foundation, the Lucie Burgers Foundation, the International Primate Protection League, and Karl Ammann provided the financial backing needed to carry out the research. Hans Wasmoeth, Sunny Kortz, Jan van Hooff, Laura Darby, Adam Singh, John and Terese Hart, Hans Breeuwer, Guido van Reenen, Wendy Atkins, Kisangola Polycarpe, Maurice Lilongo, Beth Taylor, Thurston and Kitty Hicks, and Vincent Nijman each made invaluable contributions. Ron Pontier and Dan Stears of Africa Inland Missions flew us into Bili. Finally, *merci mingi* to our skilled team of field assistants: Ligada Faustin, Chief Mbolibie of Baday, Olivier Esokeli, Seba Koya, Dido Makeima, Jean-Marie Masumbuko, Benoit Imasanga, Makassi, Likambo, Likongo, Garavura, and Kongonyesi.

Chloé Cipolletta would like to express her most sincere gratitude to the trackers of the Primate Habituation Program: their knowledge and skills were integral to the success of the program and their endless good humor allowed us to cope during the early, most difficult stages. Thank you again for all you taught me, *singila mingi.* I am grateful to Martha Robbins and Christophe Boesch for providing the opportunity to share Mlima's story in this book and especially to Martha, for kindly offering much needed editing support of the various versions of the manuscript. I was immensely fortunate to have met David Greer during my first year at Dzanga-Sangha: there is no question that I would not have stayed so long in the forest and kept my morale up and my motivation going if I had not had David's support throughout. Many thanks to the Dzanga-Sangha Project staff and colleagues, and special thanks to the assistants, volunteers, and researchers who greatly contributed to the Primate Habituation Program and with whom I shared my life for nine beautiful years. Finally, I would like to thank my father for always encouraging me to follow my passion, for supporting me in hard times and for sharing this experience with me in the forest and in the heart: *grazie mille!*

Jojo Head would like to thank first and foremost the Agence nationale des parcs nationaux and the Centre national de la recherche scientifique et technique of Gabon for permission to conduct our research in Loango National Park. Thanks also go to Loïc Makaga, Luisa Rabanal, and Erick Reteno for being the best colleagues anyone could ask for in the field. We are particularly grateful to Mr. R. Swanborn for his continued financial and logistical support of the Loango Ape Project. Thanks to Christophe Boesch and Martha Robbins for their excellent supervision and dedication to the apes. And of course the biggest thanks must go to the gorillas and chimpanzees of Loango National Park, who allowed me into their forest and let me share their world.

Crickette Sanz and David Morgan are deeply appreciative of the opportunity to work in the Nouabalé-Ndoki National Park and especially the Goualougo Triangle. This work would not be possible without the continued support of the Ministry of Science and Technology of the Republic of the Congo, the Ministry of Forest Economy of the Republic of the Congo, and the Wildlife Conservation Society's Congo Program. We would also like to recognize the tireless dedication of Jean Robert Onononga, Crepin Eyana-Ayina, Sydney Ndolo, Abel Nzeheke, Wen Mayoukou, Igor Singono, Marcel Meguessa, and the Goualougo tracking team. Special thanks are due to Mike Fay, Bryan Curran, Fiona Maisels, Paul Elkan, Sarah Elkan, Emma Stokes, Mark Gately, Bourges Djoni, Pierre Ngouembe, and Domingos Dos Santos. Our chapter was greatly improved by comments from Martha Robbins and Christophe Boesch. Grateful acknowledgment of funding is due to the U.S. Fish and Wildlife Service, National Geographic Society, Wildlife Conservation Society, Columbus Zoological Park, Brevard Zoological Park, and the Ape Conservation Effort.

Martha Robbins thanks the Uganda Wildlife Authority and Uganda National Council for Science and Technology for permission and support for her project in Bwindi Impenetrable National Park, Uganda. Research with the Kyagurilo Group was facilitated through the Institute of Tropical Forest Conservation (ITFC) and Wildlife Conservation Society. Financial support from the Leakey Foundation, U.S. Fish and Wildlife Service, National Geographic Society, and especially the Max Planck Society is gratefully acknowledged. Thanks go to all the field assistants of ITFC for their dedicated work and data collection with the gorillas, especially Tibenda Emmanuel, Twinomujuni Gaad, Mbabazi Richard, Ngamganeza Caleb, Kyamuhangi Narsis, Byaruhanga Gervasio, Twebaze Deo, Mayooba Godfrey, Tumuhimbise Ambrose, Musinguzi Dennis, Tumwesigye Philimon, and Murembe Erinerico. Special thanks are also due to Nick Parker, Sarah Sawyer, Joel Glick, Nicole Seiler, Maryke Gray, and Alastair McNeilage for their assistance with the Bwindi gorilla research project.

We thank Christophe Boesch, Thomas Breuer, Caroline Deimel, Barbara Fruth, Cleve Hicks, Shelly Masi, Sonja Metzger, Ian Nichols, Martha Robbins, and Crickette Sanz for allowing us to reproduce their photographs; full credit for each will be found on page 175.

The production of this book was only possible through the assistance of several people. We thank Melissa Remis, Andrew Robbins, and especially Kelly Stewart for their enthusiasm and support of the project, as well as for their suggestions for improving the text. Mimi Arandjelovic, Claudia Nebel, and Jojo Head offered useful opinions on the photographs. Lastly, we'd like to thank Jenny Wapner, Lynn Meinhardt, and Rose Vekony at the University of California Press for their enthusiasm, efficiency, and patience in ensuring that this book went to press.

Introduction

Who, What, Where, and Why

MARTHA M. ROBBINS

People's fascination with chimpanzees, gorillas, and bonobos and their African habitat is undisputed. Chimpanzees are known for their intelligence, bonobos for their seemingly peaceful nature and highly sexed behavior, and gorillas for their magnificent size and strength. Additionally, people are intrigued, not only by the mystique of the wild apes, but also by the difficulties faced by field-workers living in Africa. One benefit of lots of traveling that I do as part of my job is that I meet many people in the United States, Europe, and Africa. When I explain that I study gorillas, I am typically met with a small handful of responses: "Oh, is that like what that lady, what is her name—Jane Goodall or Dian Fossey—does?"; "Hmmm, I've never met anyone who does that kind of work. It must be different from being an accountant"; "Whoa, you must be crazy!" While I'm always happy that people are curiously interested, I am also continually struck by how little most people know about the great apes. I'm concerned by this for two reasons. First, we can learn much about ourselves and our world by understanding the great apes, our closest living relatives. Second, the apes are all at risk of extinction, and the first step in preventing their decline is for more people to understand them.

This book is about discovering African apes. As a group of scientists and conservationists working with African apes, we decided to put together these stories and photographs so that people will get excited about chimpanzees, bonobos, and gorillas in their natural habitats and their conservation. We wanted to write stories that combined our first-hand experiences of research, conservation, and adventure. This book contains narratives about individual chimpanzees, gorillas, and bonobos and the people who observe them. We have discovered apes on many different levels: their individual life histories and personalities, their social lives within groups, their ability to live in the harsh environments they inhabit, and the reality of the threats they face. Field biologists studying wild apes are in the unique position of spending vast amounts of time in the forest getting to know their study subjects up close and personally.

While the focus of our work may be to address scientifically relevant questions, by the very nature of the work in African forests, we also deal on a daily basis with the threats that put apes at high risk of extinction, and we have experiences that by most people's standards would be considered adventures. The narratives in this book will describe different aspects of the social world of the great apes through the eyes of the researchers studying them. We hope that true stories of individual animals and situations will draw special attention to their fate. Nearly all the narratives offer details on at least one individual ape—both in terms of its individual social situation as well as the circumstances that are beyond his/her control due to the destructive forces of humankind. This book is not meant to be a comprehensive guide to all African ape field sites, nor does it contain everything we know or discuss all the issues relating to great ape behavior, ecology, and conservation.

But before we delve into the narratives, it is necessary to give some background to put everything in context, especially for those readers who don't know much about great apes. More specifically: What makes an ape an ape? Who are the apes? Where do apes live? How do we do our work? And last, but perhaps most important, why is this work vital? Next, I'll address all these questions except for the "how," which is covered in chapter 1. This introduction is meant to provide such background, so if you would like to simply get to the stories, skip it and move on.

APES: WHAT, WHO, AND WHERE?

Chimpanzees, bonobos, gorillas, and orangutans are all great apes. As primates, they have binocular vision, grasping hands and feet, and proportionally larger brains relative to their bodies than other mammals. There are about 350 living species of primate found in Africa, South America, and Asia, most of them in the tropics, but a few in more temperate environments.

The great apes are distinct in several ways, physically and genetically, from other primates (e.g., monkeys). Anthropologists devote a great deal of energy to analyzing differences in skeletal and genetic characteristics, but I won't go into great detail here, because doing so quickly becomes rather technical and laden with specialized terminology. The great apes are much bigger than other primates and, most notably, they do not have tails. The brains of great apes are proportionally larger relative to their bodies than those of monkeys. They also have a distinctive way of walking called "knuckle-walking," which means exactly what it sounds like. As with all primates except humans, they walk quadrupedally (on both their hands and feet), but they place only the knuckles of the hands on the ground, not the entire hand.

Evolutionarily, the great apes branched off from the other primates about 16 million years ago, as shown in the phylogenetic tree on the next page. There were many other species of great apes living in the past, but unfortunately the fossil record is relatively weak for this group of primates in Africa from 5–10 million years ago, so we know very little about them. Today few species remain.

Orangutans are the only great apes currently living outside of Africa. They are found in Sumatra and Borneo, in Asia. Chimpanzees, bonobos, and gorillas all live within 500 kilometers to the north and south

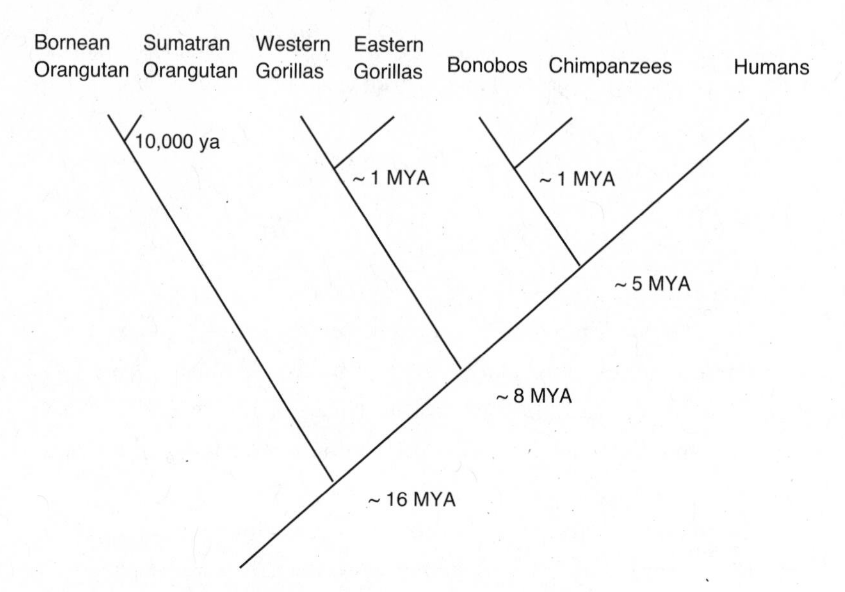

Figure 1. Great ape phylogenetic tree.

of the equator, spanning west to east from Senegal to Uganda (see map, opposite).

Of these African species, chimpanzees occupy the largest area, because they can tolerate a much greater variety of habitats than either bonobos or gorillas. The majority of chimpanzees live in tropical rain forests, but they can also survive in drier habitats known as savanna woodlands.

Chimpanzees and bonobos are closely related and to the uninitiated observer look quite similar, but they are different species. Bonobos have seemingly longer limbs, and their faces look quite different as well. Behaviorally, bonobos have totally different, high-pitched voices, which sound more like those of birds than those of chimpanzees. Sex appears to be much more important in bonobo society than among chimpanzees. Bonobos are restricted to a large area south of the Congo River in the Democratic Republic of the Congo (DRC), and they are never found living together with either chimpanzees or gorillas.

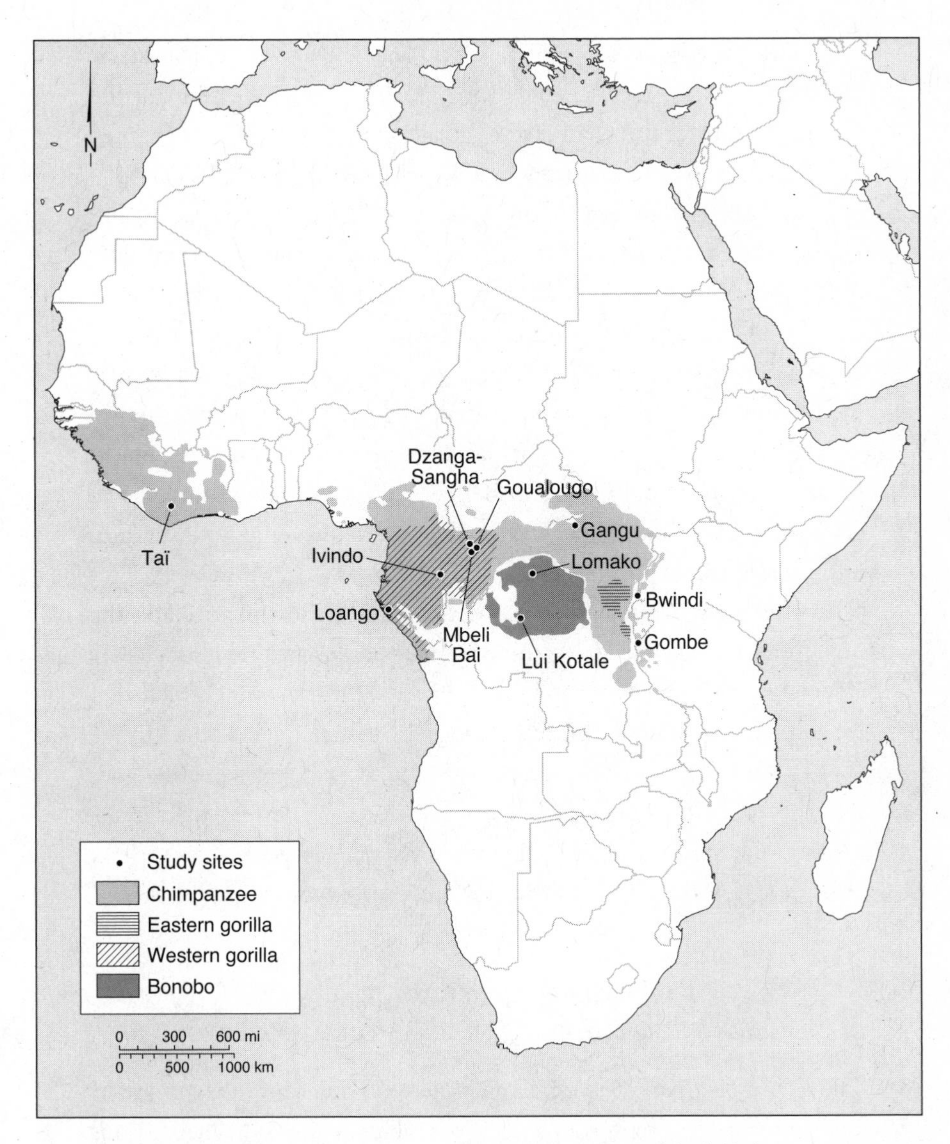

Map 1. Distribution of the great apes in Africa and locations of the field sites discussed in this book.

Currently, gorillas are split into two species, western and eastern, based on physical, genetic, and ecological differences. Western gorillas have much shorter hair than eastern gorillas, which is most likely related to differences in temperature where they live. Eastern gorillas are primarily black, with the exception of the silver hair on the backs of adult males (hence the term "silverback"), whereas western gorillas typically have a patch of red hair on their heads, and their bodies sometimes look browner. As the names imply, western gorillas live in the western part of central Africa, in areas also inhabited by chimpanzees. Eastern gorillas are divided into two subspecies, Grauer's gorillas, which are found in eastern DRC, and mountain gorillas, found only in two populations in Rwanda, Uganda, and the DRC. Between the western and the eastern gorillas, in a large block of central Africa, stretching as far as the eastern DRC, there are no gorillas.

In this book, we focus on the African apes, and thus exclude the orangutans. We decided to do this primarily because we are most closely associated with people who work in Africa. However, we also felt that we could create a more cohesive package by restricting ourselves to the apes in Africa. Nonetheless, orangutans are extremely fascinating apes in terms of their ecology, sociality, and culture. Moreover, their numbers are declining at an alarming rate. We encourage readers to look elsewhere to learn about orangutans (see, e.g., www.orangutan.org).

WHAT ARE THE BEHAVIORAL AND ECOLOGICAL DIFFERENCES AMONG THE APES?

The African apes exhibit vast diversity in their diets, how they use their environments, and their social lives (see table on pp. 8–9). There also is a great deal of variation in the behavior of each species of ape, depending on where they live. Here I give an overview of their behavior to lay out some of the most obvious differences among chimpanzees, bonobos, and gorillas. However, one of the fascinating aspects of studying great apes in different locations is discovering how much variation there is, sometimes even in locations that are very near one another.

Chimpanzees and bonobos have a diet that largely consists of fruit, but also includes leaves and other plants, as well as meat. They will sometimes hunt monkeys or duikers (small antelope). In contrast, and despite their huge size, gorillas are strictly vegetarian, except for some consumption of insects, such as ants and termites. Gorillas also eat fruit if it is available, but they rely much more heavily on vegetation such as leaves or herbs growing in the forest understory. Mountain gorillas, which live in a high-altitude environment, with little or almost no fruit, almost exclusively eat herbaceous vegetation.

Both chimpanzees and bonobos live in very large groups, typically of over fifty individuals, including many adult males and females. They inhabit what is called a fission-fusion society, meaning that group members are not always in close proximity. In fact, it is extremely rare to find most or all group members together. Instead, they form smaller subgroups or "parties," usually of two to ten individuals, with constant changes in who is with whom; for example, some will join one another, or fuse, while others will split off, or fission. This pattern of fluid subgrouping is presumably a way to reduce competition within the group. Adult females typically are with their weaned but not yet adult offspring as well as their younger offspring.

Despite the similarities in group size and structure between chimpanzees and bonobos, there are large differences in how the sexes of the two species interact with each other. In chimpanzees, the males are clearly dominant over females. In bonobos, there is no clear pattern during one-on-one interactions of whether males or females are the dominant sex. However, females frequently form alliances with one another and these coalitions are successful in out-competing and dominating individual males, so in general bonobo females can be considered to be dominant over the males. Conversely, chimpanzee males form strong alliances with one another, and in most situations, although

Comparison of the Main Aspects of Ecology and Behavior in the African Great Apes

	Mountain Gorilla (*Gorilla beringei beringei*)	Western Gorilla (*Gorilla gorilla*)	Bonobo (*Pan paniscus*)	Chimpanzee (*Pan troglodytes*)
Group composition	One-male or multi-male, multi-female; males may be solitary	One-male, multi-female; males may be solitary	Multi-male, multi-female	Multi-male, multi-female
Group size	~2–60	~2–20	~30–150	~30–150
Grouping pattern	Cohesive	Cohesive	Fission-fusion; parties tend to contain more females than males	Fission-fusion; parties tend to contain more males than females
Dispersal patterns	Both sexes disperse or remain in natal group	Both males and females disperse	Females disperse; males remain in natal group	Females disperse; males remain in natal group
Habitat	Afro-montane rain forest (1,500–3,700 meters altitude)	Tropical lowland rain forest	Tropical lowland rain forest	Tropical rain forest and savanna woodland
Diet	Predominantly herbaceous vegetation; some fruit, when available; occasionally ants	More fruit than mountain gorillas (but still rely heavily on leaves and herbs); termites and ants	Predominantly fruit, but also herbs and leaves; termites and ants; occasionally meat	Predominantly fruit, but also herbs and leaves; termites and ants; occasionally meat

Female-female relationships	Weak dominance hierarchies; weak affiliative bonds	Weak dominance hierarchies; weak affiliative bonds	Unknown	Weak dominance hierarchies, sometimes strong affiliative bonds
Male-male relationships	Strong dominance hierarchies in multi-male groups; highly competitive; almost no affiliative interactions	Adult males almost never co-reside	Strong dominance hierarchies; lack of alliances and affiliative bonds	Strong dominance hierarchies, but also strong affiliative bonds and alliances
Male-female relationships	Male dominance, but strong affiliative relationships	Male dominance, but strong affiliative relationships	Females typically dominate males through the use of female alliances. Inconsistent pattern of dominance in one-on-one interactions; strong affiliative bonds, especially between mothers and sons	Males dominant over females; variability in strength of affiliative bonds between populations
Habitat utilization	Overlapping home ranges	Overlapping home ranges	Overlapping home ranges	Territorial

not all, females have relatively weak social bonds with one another. Relationships among females are much stronger in bonobos than among chimpanzees, and we are still learning about interactions among male bonobos. Bonobos have been studied much less in the wild than chimpanzees, and understanding their social relationships is currently one of the most interesting areas of research. Only recently, field studies have shown that male bonobos form strict dominance hierarchies, and we still do not know if dominance relationships exist among female bonobos.

Gorillas live in smaller, much more cohesive groups than chimpanzees and bonobos. Groups usually consist of 10–20 individuals, but they grow to be as large as 30–60 in the Virunga Volcanoes region. All members of a group are almost always within about 100–200 meters of one another. While most groups of western gorillas contain usually only one adult male, mountain gorilla groups often contain two. This variability in the number of adult males among different populations of gorillas has implications for the types of social relationships in the groups, as well as for the stability and longevity of groups. One common feature of gorilla groups is that the dominant silverback reigns over all other group members, which isn't surprising when you consider how much larger adult males are than adult females. The strongest social relationships exist between the silverback and his females, in contrast to relatively weak social relationships among the adult females.

One of the reasons that social relationships among the sexes vary is due to differences in which sex remains in the natal group (i.e., the group into which born) and which emigrates out. One shared feature among all the African apes is that females typically migrate out of their natal group shortly before reaching maturity, thus reducing inbreeding. In chimpanzees and bonobos, all males stay in their natal group, which is one reason why male chimpanzees have such strong social bonds with one another. However, bonobo males tend to retain very strong relationships with their mothers, and it is not known why they don't form strong alliances. Gorillas are different in that males tend to migrate out of their natal group; nearly all male western gorillas leave the natal group, and about half of male mountain gorillas do. These emigrant males are then solitary, unless they are successful in attracting females to form a new group, which can take years.

The African apes also differ in how they defend the areas of the forest that they use. Chimpanzees are highly territorial, regularly performing "boundary patrols," and can be extremely aggressive toward members of neighboring communities. It is believed that this is because of the need to protect their females and the food resources necessary to maintain a strong social group. Bonobos appear to have home ranges that overlap with other groups, and they will be aggressive toward their neighbors, but they are not territorial like chimpanzees. Gorillas do not defend territories. Instead, they have "home ranges," which may overlap among neighboring groups, but this doesn't mean that everything is calm and peaceful when two groups meet. Males will go to great lengths to show their strength and prowess in attempts to attract new females into their groups and to retain the females they already have.

The stories in this book span several different field sites of the four African ape species (chimpanzees, bonobos, western gorillas, and mountain gorillas) and illustrate their diversity. You may ask why it is important to study apes in many different locations. If it is so much work and effort, why not study each species in only one location? Undoubtedly, nobody would think that a successful way to totally understand humans would be to study them in only one location, such as New York City or the Kalahari Desert. The habitats where each species of African ape lives vary in many regards. Additionally, we know that chimpanzees and bonobos exhibit "culture," or variations in behavioral patterns that cannot be attributed to either genetic or ecological differences. For example, the techniques used to extract ants, a good source of protein, from their nests are different in Côte d'Ivoire and Tanzania. The chimpanzees in Côte d'Ivoire also use stones as hammers to crack open nuts, which is not observed in any eastern African chimpanzees. Gorillas most likely also exhibit "social traditions" or variation in particular behaviors, but to date they have been studied intensively in fewer locations than chimpanzees.

Although some of the stories in this book show the apes in a very aggressive light, they are not included to shock readers or to portray any of the apes as being exceedingly aggressive. Rather, the authors chose incidents that clearly stuck out in their minds as influencing how they viewed the apes that they spent a great deal of time studying. Highly aggressive behavior in the African apes is very rare, but it clearly has an impact on their societies. The daily routine of most humans is less interesting than rare incidents such as conflicts, and the same is true with the apes. As with humans, most interactions among the apes are peaceful and calm.

In many places in the book, we have been shamelessly anthropomorphic—attributing human qualities to animals. As scientists, we conduct studies from an ecological and evolutionary perspective and do not view our study subjects in the same way we view humans. Yet it is impossible to observe wild apes and not see certain parallels with human emotions and personalities. In describing the apes in this fashion, we are taking this opportunity to show their more "human" side, as well as our own less scientific views of their actions.

WHY SHOULD WE BE INTERESTED IN THE AFRICAN APES?

Why do humans behave as we do and live in the types of societies that we do? People around the world live in some variety of families, communities, and societies. We have alliances, friends, family, acquaintances, and enemies. Premodern societies hunted and gathered food and eventually developed agriculture. We have invented tools as simple as forks and as complex as computers and airplanes. We communicate not only through intricate languages but with gestures such as pointing and smiling. And most fascinating, there is huge variation in nearly every aspect of human behavior. How and why did this come to be?

Obviously, we can learn a great deal about human behavior by studying ourselves. However, to most completely understand human behavior and our origins, it is useful to study the species most closely related evolutionarily to humans, which are the African apes. Following evolutionary theory, scientists assume that the environment in which animals live shapes their lives: what they eat, where they go, and how they interact with one another. A feedback system develops in which animals may shape their own environment, with humans doing so to an extreme. Many people, including some anthropologists, will go to great lengths to emphasize the differences between humans, apes, and other animals. We are certainly not claiming that humans are just the same as chimpanzees or gorillas, but, as already noted, after spending even a little time watching great apes, most people see some striking similarities. Furthermore, many behaviors that were initially considered uniquely human, and therefore could be used to define the "human," have been found to exist in other animals, particularly chimpanzees. For example, tool use, culture, cooperative hunting, and warfare are all observed in one form or another in wild chimpanzees. Comparing and contrasting ourselves with the African apes can help us understand not only some our most horrendous behavior, such as warfare, but also our most intriguing behavior, such as language and culture.

A second, but perhaps more urgent, reason to be interested in African apes is that they all are at risk of extinction. It is not inconceivable that no apes, or shockingly few, might remain in the wilds of Africa before the end of this century. At the most extreme, only about 800 mountain gorillas live in two small, isolated forests in Uganda, Rwanda, and the Democratic Republic of the Congo. Many more chimpanzees, bonobos, and western gorillas are living in central and west African forests, but it is not possible to say how many, because we don't exactly know. One of the big challenges facing biologists today is developing accurate methods for estimating the sizes of ape populations that can be used in a cost-effective and timely manner so that we can monitor changes in their numbers and the effectiveness of different conservation strategies. Despite not knowing exactly how many apes there are in Africa, all the evidence suggests that their numbers are declining at an alarming rate due to deforestation and mining, hunting, and diseases.

As the number of humans on earth continues to grow, there is increasing demand for natural resources, and Africa is one of the remaining places where timber and minerals can be found in large supply. Companies extract timber, oil, coal, and minerals throughout central Africa, including many locations where great apes are found. Large-scale logging and deforestation are less prevalent in eastern Africa, largely because so few large tracts of rain forest remain. However, habitat degradation can still occur on a smaller scale when local people gather firewood as fuel for cooking.

Another huge threat to great apes is hunting, or what is called "the bushmeat trade." In many regions of Africa, livestock is not kept as a source of meat; instead, people eat what is hunted out of the rain forest. Most of the animals killed and eaten are small antelope and monkeys, but gorillas, chimpanzees, and bonobos are eaten as well. Westerners are typically disgusted by the idea of eating a great ape, and many people in eastern Africa have similar taboos, yet in many locations they are simply meat like any other animal. Even though fewer great apes are killed than other animals, killing even a few individuals annually has a devastating effect on a population, because their reproductive rates (time until first offspring is produced, time between having offspring, etc.) are so low. Additionally, the bushmeat trade is amplified in areas where logging or mining occurs. When companies open up networks of roads to get timber or other materials out of the forest, this increases the amount of traffic from remote areas to towns and cities and makes it easy to transport bushmeat, which encourages illegal hunting.

Diseases, both those naturally occurring in their environments and others transmitted by humans, are the third major threat facing the African apes. Understanding the impact of disease on wildlife, including apes, has only gained the attention of biologists in the past decade or so. Perhaps one of the most dramatic cases of disease impacting wildlife is Ebola hemorrhagic fever (EHF). There is strong evidence indicating that Ebola has swept across much of central Africa in the past decades, decimating populations of gorillas and chimpanzees. In one location in the Republic of the Congo, scientists were able to document that 90 percent of the gorilla population had died in a matter of months. Currently, there is no vaccine or cure for Ebola, which has serious implications for both the humans and apes living in these areas.

As increasing numbers of humans move closer to ape habitat and ecotourism expands as a conservation strategy, the risk of transmitting disease from humans to apes rises. The great apes do not have the same immunity to certain respiratory and intestinal diseases that we have, nor can they easily be vaccinated or treated if they become ill. In recent years it has been shown that humans have transmitted respiratory disease to chimpanzees that may manifest as a common cold in humans, but is deadly in the apes. While the risk of disease transmission can be reduced by taking precautions when in close contact with the apes, it may be difficult to control as the ape-human interface includes more and more people living near the apes.

Such details of the threats facing the great apes are most likely nothing new to most readers, since the general public is constantly bombarded with depressing statistics and media reports of animals facing extinction, to the point that most people become numbed to them and feel that nothing can be done. Yet these threats are real, and in a few decades we may be telling children about great apes only through books and films, with none left in the wild. Many conservation organizations and governments are working diligently to save the African apes through a wide variety of methods. The most common approaches to conservation include law enforcement, education, research, community development projects, and ecotourism. None of these approaches in and of themselves are particularly complex, yet using these methods in an effective, multifaceted way so as to prevent ape populations from declining is remarkably complex in areas suffering from poverty, corrupt governments, and lack of good schools or the financial capital for development. The resources available for conservation in Africa are highly inadequate. It may sound melodramatic, but we are currently attempting to stop what amounts to a raging fire by splashing a few cupfuls of water on it.

There are definitely times when the threats and problems seem insurmountable, but we cannot give up hope for the future. There are big and small ways in which everyone can help. All the contributors to this book are passionate about their profession and the apes, and we want to share this passion with others to spur interest in them and their conservation. The discoveries, adventures, and joys of studying the apes far outweigh the aggravations, illnesses, frustrations, and hardships. We hope that this is conveyed through our stories and photographs. However, we also very much wish to convey a sense of the emergency the apes are facing as their forests are destroyed and their numbers dwindle.

FURTHER READING

Boesch, Christophe. 2009. *The real chimpanzee: Sex strategies in the forest.* Cambridge: Cambridge University Press.

Boesch, Christophe, and Hedwige Boesch-Achermann. 2000. *The chimpanzees of the Taï Forest: Behavioural ecology and evolution.* Oxford: Oxford University Press.

Boyd, Robert, and Joan B. Silk. 2008. *How humans evolved.* 5th ed. New York: Norton.

Caldecott, Julian, and Lera Miles, eds. 2005. *The world atlas of great apes and their conservation.* Berkeley: University of California Press.

Campbell, C.J., A. Fuentes, K.C. MacKinnon, S. Bearder, and R. Stumpf, eds. 2010. *Primates in perspective.* 2nd ed. Oxford: Oxford University Press.

Fossey, Dian. 1983. *Gorillas in the mist.* Boston: Houghton Mifflin.

Goodall, Jane. 1986. *The chimpanzees of Gombe: Patterns of behavior.* Cambridge, MA: Harvard University Press, Belknap Press.

Harcourt, A.H., and K.J. Stewart. 2007. *Gorilla society: Conflict, compromise, and cooperation between the sexes.* Chicago: University of Chicago Press.

Schaik, Carel P. van. 2004. *Among orangutans: Red apes and the rise of human culture.* Cambridge, MA: Harvard University Press, Belknap Press.

Strier, Karen B. 2006. *Primate behavioral ecology.* 3rd ed. Boston: Allyn & Bacon.

Waal, F.B.M. de, and Frans Lanting. 1998. *Bonobo: The forgotten ape.* Berkeley: University of California Press.

Plate 1. An adult male chimpanzee in Taï National Park, Côte d'Ivoire.
Plate 2. An adult female bonobo in Lui Kotale, Democratic Republic of the Congo.

Plate 3. A juvenile chimpanzee peers down curiously at observers in the Gangu Forest. *Plate 4*. An adult male chimpanzee watching observers.

Plate 5. An adult female bonobo resting among vines.
Plate 6. An infant bonobo playing on his mother.

Plate 7. Bonobos feeding together in a tree. *Plate 8*. The attention of a mother, a juvenile, and an infant bonobo caught by another ape higher in the tree.

Plate 9. An adult male bonobo reclining in a tree.
Plate 10. Grooming among female bonobos.

Plate 11. An intense chimpanzee grooming session.
Plate 12. Infant chimpanzees investigating a flower.

Plate 13. Dorothy and her infant in Goualougo Forest.
Plate 14. Tool use: Dorothy uses a stick to extract honey from a bee nest.

Plate 15. A group of western gorillas in Mbeli Bai. *Plate 16*. An adult female western gorilla wading through Mbeli Bai to feed on vegetation.

Plate 17. An infant western gorilla stretching and yawning on his mother's back as she feeds. *Plate 18*. A knuckle print left behind in the mud, a crucial sign in tracking western gorillas.

Plate 19. An infant western gorilla.
Plate 20. A playful infant western gorilla in Mbeli Bai.

Plate 21. An adult female mountain gorilla and an infant huddle together during a chilly downpour. *Plate 22*. A blackback male mountain gorilla. At ten to twelve years of age, the males still lack the large crest on their heads.

Plate 23. An adult female mountain gorilla with a newborn infant.
Plate 24. Rukina removes some vegetation stuck in his teeth.

Plate 25. Rukina rests with his family.
Plate 26. Rukina grooms an adult female, while his first son plays on his back.

Plate 27. Researchers measuring tools used by chimpanzees in Loango National Park, Gabon. *Plate 28*. The Goualougo Chimpanzee Project camp, in Republic of the Congo.

Plate 29. Researchers watching chimpanzees in the Goualougo Triangle, Republic of the Congo. *Plate 30*. Lui Kotale research camp, Democratic Republic of the Congo.

Plate 31. Rukina in his prime.

Plate 32. A silverback western gorilla walking in the forest.

ONE

Discovering Apes

MARTHA M. ROBBINS

HOW DO WE STUDY APES IN THE WILD?

Being a field biologist is often perceived as much easier and more glamorous than it actually is. Some people may even say that it is not "a real job." The image is of someone in khakis and safari hat confidently striding through the rain forest, notebook in hand, stopping frequently to glance through binoculars and jot down every move of the nearby apes. While we write down lots of things, we certainly don't write down everything we see. As with all scientists, we test particular hypotheses or predictions and use systematic methods and protocols that are designed to provide specific data. We then summarize our findings and write articles for journals that undergo scrutiny by our colleagues. In this way, we not only expand our knowledge but explain unexpected findings, which typically lead to even more questions and more research.

The challenge of fieldwork is that, as opposed to work in a laboratory, we cannot control, nor do we want to control, the natural environment. We want to observe the apes as themselves and have no influence, or as little as possible, on what they are doing. This poses the first difficulty: how do you obtain "natural, undisturbed" observations of wild animals that are inherently afraid of humans?

The main method used to get detailed observations on wild apes is to "habituate" them. What this means is that through repeated neutral contact with human observers, the apes gradually lose their fear of humans and will accept us watching them from close distances. Once habituated, the apes are still wild. We don't provide food for them or create any situation where they are reliant on humans for anything. They are not captive, nor do they become pets. Habituating some primates can take only a matter of months, but it takes years to habituate apes. It may take only one to two years to habituate mountain gorillas, whereas it can take five or ten years to habituate western gorillas, bonobos, or chimpanzees.

There are several reasons why it takes so long to habituate apes. The first challenge is simply finding the apes. If you aren't encountering any apes, you can't begin to habituate them. Gorillas and chimpanzees live at relatively low densities of only one to four individuals per square kilometer. Despite their large size, this translates into a low probability of simply bumping into them if you are walking through the forest. Therefore we rely on a variety of detective-work-style approaches. We look for any visual or auditory signs they may leave in the forest. In the case of chimpanzees, who regularly make very loud calls, it is useful to sit quietly in an area where you think they are, wait until you hear them call, and then try to find them. For gorillas, which are heavier animals than chimpanzees and spend more of their time foraging and resting on the ground, it is possible to see footprints and other signs, such as remains of plants that they have eaten or their feces. Then you can "track" where they have moved through the forest, if you are skilled enough, given that the signs are often extremely subtle and difficult to see. Mountain gorillas live in forests where there is a dense understory of herbs and shrubs, so it is relatively easy to see where they have gone. This makes them much easier to track and contributes to why it is easier to habituate them. In contrast, western gorillas live in forests with little understory, so it is much more difficult to track them. In sum, it takes a tremendous amount of patience. Researchers often spend weeks or months searching for signs and getting only rare glimpses of a chimpanzee or gorilla.

Once you find the apes, the next step is to "convince" them that we mean no harm and that we are simply neutral factors in their environment. If the apes fear humans, however, they typically flee immediately upon seeing any observers. It can take months and months to see a change in behavior, in which a gorilla or a chimpanzee will wait a few minutes before running away, and then even more time before they will resume their normal behavior with people nearby. Because chimpanzees and bonobos live in fission-fusion societies, even though you may be meeting the same group daily, it is unlikely that you are seeing the same individuals every day, whereas with gorillas, which live in cohesive groups, you will meet the same individuals daily. This is one reason why it takes longer to habituate chimpanzees and bonobos than gorillas. Yet another issue to consider with gorillas is that if they feel threatened at a close distance, the silverback male will defend himself and his group by charging. It takes nerves of steel (and perhaps a little bit of insanity) to hold your ground and not run when a four-hundred-pound silverback is screaming and running directly toward you. However, these charges are primarily displays, because given the risk of getting hurt themselves, most animals do not actually want to attack. In any case, apes slowly become more accepting of humans, eventually developing "trust" of their human observers and allowing us to watch their lives.

Needless to say there are some risks to habituating great apes, and it is not an endeavor to be taken lightly. The very process of habituating wild apes causes them a great deal of stress and fear. By eliminating their fear of humans, we are, moreover, exposing habituated apes to the risk of poachers approaching and easily killing them. Therefore it is necessary to provide constant monitoring and protection to habituated apes. It is a long-term commitment. In addition, by coming in close contact with the apes, we are exposing them to diseases, particularly respiratory and intestinal ones, that they would not normally encounter or have resistance to. Old photographs of Dian Fossey grooming a gorilla may inspire many young scientists and tourists to want to do the same, but such interactions could lead to the gorilla dying. In the past decade or so, nearly all field sites have established health guidelines that include (to name a few) requiring all visitors to be vaccinated against many diseases, not allowing people to go into the forest if they are ill, following various hygiene procedures, and wearing surgical masks when near the apes. Some people may feel that these risks and the negative aspects of habituation do not justify ever doing it. However, all of the contributors to this book, as well as many other scientists and conservationists, believe that the benefits of being able to observe wild apes closely far outweigh the costs. For example, several studies have shown that the presence of researchers and/or ecotourism programs reduces poaching and other illegal activities. It is very difficult to get people interested in animals that they can't see or don't find attractive. By learning details of the social lives of great apes, people come to see the value of protecting them.

In addition to habituation, scientists use other indirect methods to study the great apes. We can gather a great deal of information about the apes' behavior by studying their forest and the remains that they leave behind. For example, much can be deduced about tool use by examining what look like sticks and rocks to the uninitiated, but in fact are brushes, probes, and hammers left behind by clever apes.

An unusual method of studying western gorillas involves sitting on a platform in a large clearing (called a *bai*) where the animals regularly come to feed on aquatic vegetation (see chapter 8). This method has the disadvantage of observations being possible only when the gorillas decide to go to the clearing, which may be as infrequently as once a month or less. The main advantage is that without habituation, if you have chosen a good bai, it is possible to monitor the group composition of ten or more social groups over the course of a year, which would take phenomenal effort through habituation.

To get the best understanding of ape ecology, in addition to studying what the apes eat and where they range, it is necessary to know what is actually in their environment. To do this, we are forced to become botanists and conduct detailed studies of food availability. This involves spending months and months counting and measuring the size of all herbs, shrubs, and trees in plots that represent only a small proportion of the forest. Because fruit is not found on trees all the time, we also set up "phenology studies," where we routinely check known individual trees for the presence of flowers, fruits, and young leaves. Typically, study sites monitor from 300 to 1,000 trees at least once a month.

The most unattractive item that apes leave behind actually provides a wealth of information: their dung. By sifting through feces, it is possible to learn a lot about what the animals have eaten. For example, if a gorilla eats fruit but doesn't crush the seeds, they are intact in its feces, and then we know which plant species it has eaten. Additionally, cells from the intestinal lining of an animal are excreted, making it possible to extract DNA and conduct genetic analysis from feces. This enables us to answer many questions that only a few decades ago were considered impossible to address with wild animals. Specifically, we can now determine the paternity of infants, as well as use genetics to know the whereabouts of individual animals that disperse out of habituated groups and are not physically seen regularly, but whose feces are occasionally found. Through genetic analysis we can also use feces to assist us in determining the number of apes in a particular location, which provides a more refined estimate than many other methods. Using feces to detect parasites has been done for decades, but new techniques are being developed to screen feces genetically for viruses and bacteria. This is opening up an entire new world of understanding the impact of disease on the apes, including the risk that humans pose to them. In an increasing number of ways, high-tech laboratory work is assisting wildlife biology.

In addition to the actual process of collecting data, there are many other unglamorous aspects of running a field site that are similar to many other jobs, albeit in a more remote, difficult environment. Nobody ever works entirely alone, so it is necessary to hire and manage local and international field assistants and students. Food and other supplies need to be purchased. Accommodation, which may be as basic as a tent or as relatively luxurious as a house, and vehicles or boats need to be maintained. Accounting for all expenses needs to be done. Obviously, money is necessary for all of this, so grant proposals and reports need to be written to keep projects operating. Simply getting to the field sites is not easy. To reach any of the field sites discussed in this book takes anywhere from one to three days of travel from the national capital by some combination of car, boat, airplane, and foot.

The stories in this book convey aspects of the romantic image of the work, and the adventure and thrill aspects are definitely a part of what keeps us motivated. On a day-to-day basis, the fieldwork can be mundane as well as challenging. It often involves spending long hours recording very detailed data on animals that can be difficult to find, let alone observe. All this is done come rain or shine, come marauding elephants or stinging nettles, come leeches or malaria. We spend years in the forest, because the apes are long-lived and not all aspects of their biology can be uncovered in a short period of time. The strict routine of scheduled work can also be interrupted at any point for a variety of unexpected reasons: poachers may come through the study site, making it necessary to notify the park authorities and remove snares; a field assistant may fall sick and need to be taken to a hospital that is several hours' drive away; a vehicle may break down, so that days are lost sitting around waiting for it to be repaired. Endless patience and flexibility are necessary.

Working in the forest is the easy part. Just as any scientist does, we spend innumerable tedious hours sitting in front of computers reviewing data, performing complex statistical analysis that reduces all the amazing beauty of the animals to a bunch of numbers, and writing very technical papers that receive thorough criticism from our peers. As conservationists, we may meet regularly with park authorities to provide advice on local issues, attend national and international conferences aimed at developing large-scale strategic plans, or develop educational materials for local schoolchildren and villagers to help them understand the value of protecting their neighbors the apes.

After hearing the challenges of habituating wild apes and working in remote locations, you may also be wondering why we don't simply study apes in captivity rather than go to all the trouble of habituation and working in difficult and isolated places. While captive settings are useful for some studies, such as experiments on the cognitive abilities of the apes, they are of limited use for understanding how animals have adapted to live in their natural environments. Who lives with whom, what they eat, and their "ecology" in a captive setting is determined by the zoo managers, not by the animals themselves. It is our hope that the explanations of the methods we've used and the stories that follow will convince you that we need to study and conserve apes in their natural, undisturbed environments. I'll end this chapter with a glimpse into what it is like to start a project with wild apes.

AN UNEXPECTED SIGHTING

"Are we wasting our time?" I wondered. What was the likelihood of finding any apes in this forest? I wasn't expecting actually to see gorillas or chimpanzees, but had dwindling hopes of even seeing signs of their presence, such as dung, torn scraps of vegetation, or footprints.

In 2003, Christophe Boesch and I were undertaking a month-long search in Gabon for a location to establish a new field site that contained both chimpanzees and gorillas. Despite decades of fieldwork at other locations in Africa, surprisingly little is known about the great apes found in the Congo Basin. We wanted a location where we could habituate the apes to our presence so that we could make detailed observations of their social behavior and ecological habits. Based on Christophe's twenty-five years of experience studying chimpanzees in Côte d'Ivoire and over a decade of my researching mountain gorillas in Uganda and Rwanda, our basic list of requirements for a suitable field site was short, but not easy to meet. We wanted a place that was not too remote, had no signs of poaching, was not under imminent threat of logging, and contained a healthy population of apes. Relatively speaking, Ivindo National Park met the first requirement. Whereas a hundred years ago, it would have taken Stanley, Livingston, or de Brazza weeks, if not months, to get from the coast to the depths of this forest, we had managed to get there with only a one-hour flight on a twenty-seater plane from the capital, Libreville, to Makokou, and then a five-hour boat ride down the Ivindo River to a scenic place to pitch our tents on its banks. Perhaps it was not as simple as arranging a picnic in a park, and I shouldn't understate what it takes to arrange such a trip, but we had been extremely fortunate with logistics. Following seemingly few emails and discussions with local conservationists and government officials, we had kindly been given a rough map of the area, information on a few places to camp, and the names of some local fishermen who would be happy to work as guides. After only one day of organizing in the frontier-style town of Makokou, voilà, we were loaded up with a ten-day supply of the fanciest nonperishable foods we could get, including such delicacies as oatmeal, rice, corned beef, and sardines, as well as plenty of energy and optimism, and we were on our way.

We had been told the Ivindo forest contained "naïve" apes—chimpanzees and gorillas that had had so little contact with humans that they would not fear us, but instead would be curious about the presence of an unknown, upright-walking ape, and thus presumably easier to habituate. Yet after a few ten-hour days of searching, we had seen few signs of apes or any large mammals, and it was clear that logging companies and poachers had discovered this forest before us. Machete cuts on saplings throughout the forest and the lack of bushpig were strong indications of heavy poaching. Monkeys gave alarm calls and fled immediately upon detecting us. We could see that elephants had previously frequented the forest based on the distinctive trails that they leave after decades of walking the same routes between favorite fruit trees, but we saw only a few in the flesh and some piles of old dung. The only hint of great apes was one aging pile of dung that might have been deposited by a gorilla, or perhaps by a buffalo. As we walked along a trail cut by loggers to get to the interior of the forest, I was further depressed by tree stumps of much greater diameter than my height. How much time had gone by from when those giant trees had been little seedlings until they were cut down in a matter of minutes? Presumably hundreds of years. As we passed yet more signs of poachers, I thought that the only naïve animal in this forest was us.

It was the morning of Day 6, and I was feeling pretty discouraged. It didn't help that although it was only 8 A.M., I was already hungry for lunch—hungry enough to start looking forward to a tin of oily sardines. My feet were wet because my boots hadn't dried out from a foray across a small stream the previous day. I had itchy mosquito and black-fly bites everywhere. I knew that to get through the day I should stop complaining to myself and be more positive about finding apes in this forest. Yes, it was pretty darned cool to be smack in the middle of the remote Congo Basin rain forest. Yes, the trees were huge and the forest was beautiful. Yes, maybe we'd see something interesting today . . .

After a while, Christophe and I paused to decide on a route to bushwhack through the forest. As we discussed this over the minimal map, amid the background noise of insects and birds, our conversation was abruptly interrupted by the low, distinctive hoo-hoo-hoo that could be only one animal. Chimpanzees! Based on the volume of the calls, the chimpanzees were surprisingly nearby, probably only a few hundred meters away. We were in luck—if we could locate them before they moved off.

Being the chimpanzee expert, Christophe automatically and without hesitation took a compass bearing and sprinted off in the direction of the chimp calls. I frantically ran after him. Chimpanzees can move fast and disappear instantaneously into the forest without leaving a trace. To catch up with wild chimpanzees requires a seemingly impossible combination of physical abilities: keep your eyes focused in the dim distance for quick detection of moving black shadows, keep a straight line course of the determined direction while dodging trees, and run as fast as you can over uneven ground without tripping on any vines, shrubs, or small holes, while remaining as quiet as possible, despite dry, crunchy leaves scattered everywhere on the forest floor. And then, once you get close enough for the chimpanzees to see you, you need to slow down and nonchalantly become part of the forest. After about ten minutes, as I not-so-silently-or-gracefully flailed along, Christophe suddenly froze. I screeched to a halt just behind him, and after wiping the sweat from my eyes and disentangling my binoculars, which now had a stranglehold around my neck, I quickly detected movement ahead in several places.

My first reaction was of utter amazement that we had stumbled on any chimpanzees. My second reaction, being accustomed to quiet gorillas and only a novice with chimpanzees, was a desire to ask them to turn down the volume of their deafeningly loud hooting and hollering. Yes, we could hear them and so could every living thing within a kilometer. There is no sound like it; it fills the forest and amplifies, making it seem that there are many more chimpanzees present than there actually are. Whaaa-whoo! Whaa-whooo!! WHAA-WHOOO!!! Gorillas would never be so loud. My third thought was that, okay, we have gotten a glimpse of these chimpanzees, they've started to give alarm calls, and now they are going to take off before we can see much more than the outlines of their bodies. These chimpanzees were not habituated to seeing human observers at a close distance. But instead of our quickly seeing the backsides of fleeing chimpanzees, a performance expanded before us.

Immediately, we could see that there were many chimpanzees at various levels in the trees in front, to the left, and to the right of us. A huge adult male, with the hair on his arms erect to enlarge his appearance, stared down at us from fifteen meters above in a tree directly in front of us. He was one of three adult males we could see calling at the top of their voices while watching us from the trees. After a few minutes, two of the males nimbly descended and moved off into the forest. The third male climbed higher in his tree and continued to watch us. An adult female was in a tree thirty meters to the right side of us, clutching a small baby, who peered at us with wide eyes, pink face, and comically huge ears. She watched us nervously from behind a branch, then quickly bolted from the tree and vanished into the forest, with baby attached. Nearby, to the left, a juvenile descended a tree to get a better look at us, torn between his curiosity and his apprehension. He kept throwing glances between us and his mother, who was in the tree behind him. Finally, he climbed the tree containing his mother, and together they moved away in the canopy. Another adult female with a baby was barely visible as she silently sat partially hidden behind some foliage. Next, an adult male with a distinctive bald head dramatically jumped from tree to tree, first moving closer to us, then further away. A young female with a sleek, slender body and legs watched us briefly before slowly climbing down a tree and up another one.

Without a doubt, the chimpanzees were curious about these creatures who had intruded on the privacy of their morning. Because there were so many chimpanzees and so much movement, it was difficult to decide where to look or even to try to keep track of which chimpanzee was moving where. Several of those that we initially saw ran away, but then others appeared from further behind or from high up in the trees. It was almost like the ebb and flow of the ocean, with some individuals moving closer to get a better look at us, while others edged back away from us. We counted at least twenty chimpanzees in all. All the while, the loud calls continued to echo through the forest. Eventually, the curtain fell on this drama, and each chimpanzee slowly and subtly melted into the forest.

As the last chimpanzee disappeared, the scientist in me took a look at my watch. I was surprised to see that only thirty minutes had passed since we first sighted them. It felt like an hour or two. I began to wonder whether these chimpanzees had ever seen humans before. Christophe and I didn't say anything to each other for several moments after the last chimpanzee disappeared. He generally keeps talk to a minimum in the forest, but I could sense that he was also impressed by what we had just seen. What do you say after a scene like that? Eventually, we compared notes on how many chimpanzees we had seen, their age, sex, and behavior, and then we continued on our way. I felt privileged to have seen those chimpanzees, not only because they were surviving in a very disturbed forest, but because it was such an extraordinary, extended view of so many unhabituated individuals. Yet they had given us a mere glimpse into their lives. It was almost like seeing a trailer for a good movie—you see the main characters and get a hint at the plot, but you want to see the entire show. We spent the next few days searching in the same vicinity of this chimpanzee contact, but without further luck.

Although we decided against Ivindo as a new field site, because of the high level of human disturbance in that locality, I have often wondered about those chimpanzees. Are they still out there? Poaching has undoubtedly continued in the forest. Moreover, a huge waterfall on the nearby Ivindo River was going to be dammed by a Chinese company to generate hydroelectricity. Great apes can survive in disturbed forests—but for how long and at what cost? And what about those chimpanzees as individuals? Whose peaceful morning feed had we rudely disturbed at that clump of fruiting trees? Who was friends with whom, and who felt animosity toward whom? Was the group's dominant male there? Had they come together simply to feed, or were several adult males pursuing the same adult female? How big was the community, especially given that we had seen twenty feeding together? We didn't know those chimpanzees. Although it is easy to garner concern for the threats facing the habituated apes we discuss in the following chapters, this glimpse into the lives of unhabituated chimpanzees serves as a reminder of the thousands of apes that are not habituated yet still face increasing threats to their survival.

FURTHER READING

Blom, A., C. Cipolletta, A.M.H. Brunsting, and H.H.T. Prins. 2004. Behavioral responses of gorillas to habituation in the Dzanga-Ndoki National Park, Central African Republic. *International Journal of Primatology* 25:179–96.

Cohen, J. 2010. In the shadow of Jane Goodall. *Science* 328:30–35.

Doran-Sheehy, D.M., A.M. Derby, D. Greer, and P. Mongo. 2007. Habituation of western gorillas: The process and factors that influence it. *American Journal of Primatology* 69:1354–69.

Setchell, J.M., and D.J. Curtis. 2003. *Field and laboratory methods in primatology: A practical guide.* Cambridge: Cambridge University Press.

TWO

Life and Death in the Forest

CHRISTOPHE BOESCH

CHAOS ERUPTS

It was midday in early 1998, and the sounds of the African rain forest were getting quieter, because most animals were resting. The sun was bright above us, and in front of me some black fur balls lay on the ground. They all had been resting together for an hour already. At the base of a giant tree, Lefkas slept near his mother, Loukoum, and his young baby brother, Léonardo. Mognié, his preferred playmate, was sleeping not too far away next to her mother, Mystère, and young brother, Mozart. Lefkas was seven years old, just getting more robust, but still totally under the protection of his strong mother. While Lefkas was sound asleep, Mognié dozed, with her hand playing with the incredibly small feet of her younger brother lying on the belly of Mystère. In total, fifteen of the thirty-three chimpanzees in this community were sleeping around me, and the quiet atmosphere was making me fight my heavy eyelids. But chimpanzees would not be chimpanzees if they did not surprise you even after you had observed them for seventeen years.

Suddenly, within a second, all the chimpanzees were up, and chaos erupted in the forest. A few hundred meters away some males had captured a monkey to eat. No other sound, not songbirds, insects, or even the wind, could be heard over the chimpanzees calling at the top of their voices, which are very loud. They ran to one another, embracing and kissing for reassurance—for what was still an unknown reason to me. All were very excited and, with erect hair, appeared twice their normal size. As a young male, Lefkas displayed aimlessly among his companions, mimicking the impressive displays of adult males, until he finally found Mognié. However, she was willing to be chased for no more than a few meters before turning around, and then I could see only a big, noisy, laughing ball of fur rolling on the ground as Lefkas and Mognié bit each other's toes, pulled each other's hair, and tickled each other's armpits. I watched them with a smile and thought about how far Lefkas still had to go before he would be taken seriously as the big male he wanted to be. But they were wasting time, because the rest of the chimpanzees were running noisily toward the males. So they dashed through the thick undergrowth, while I painfully tried to follow them, propelling my 5'9" frame through tangled vines, thorns, and saplings.

Chimpanzees will invest a lot of energy to get meat. Hunting small monkeys thirty meters up in the canopy is difficult enough, but it is even harder because the monkeys can easily jump across large gaps between the trees that chimpanzees are too heavy to traverse. So the solution is to work as a team, in which each of the hunters performs a complementary role. If successful, meat is a bonanza from which everyone wants a share. When high-ranking Loukoum arrived at the scene of the kill, she immediately took a central position between Macho and Marius, dominant males who were monopolizing the large adult monkey they had caught. Both were quite tolerant of Loukoum's eagerness to take pieces of meat from their share. Loukoum always was a "meat fighter" and was able to impose herself among high-ranking males to get a piece. I could still remember an incident years before when young Loukoum had fought her way to the meat, with tiny Lefkas on her belly, hitting or biting at all males that dared to limit her access, not to mention her fierce reactions to females that even thought of challenging her position. There is no doubt in my mind that females have a central position in the social network of chimpanzees, and in some particular situations, the dominant ones can easily dominate males if they want. In the Taï forest, females are more willing to challenge males for meat than in other chimpanzee populations. Loukoum was only one of the many females who had fought their way up to the highest position in the female dominance hierarchy.

During this time of meat sharing, I was not aware that Kendo, a former dominant male of the group, was Lefkas's father or that Macho was the father of Léonardo, which might have helped explain his tolerance of Loukoum. However, Loukoum often imposed herself to get meat from males that had not fathered any of her offspring. Thanks to noninvasive methods using feces, we are now able to determine family relationships in wild animals genetically, thus gaining a much better knowledge of their social relationships.

The chaos, full of screams and charges, quieted down after fifteen minutes, and everyone concentrated on eating. Suddenly, Lefkas saw that Mognié had a long piece of fur from the monkey. With envious rage, he jumped on her, and the laughing fur ball started to roll again within the group of the chimpanzees. Many were disturbed by them, and a few made some complaining grunts, but no one let themselves be distracted from what they were doing: those who had meat ate it, meatless ones begged, and Lefkas and Mognié fought over the meatless piece of fur. This piece of fur turned out to be the best pretext for play all that afternoon. Youngsters seven or eight years old sometimes seem to be interested only in playing. A youngster with as much energy as Lefkas was certainly a champion at forgetting everything if he could have a good wrestle.

THE BLACK CURTAIN FALLS?

One morning in May 1999, while working in my office in Leipzig, Germany, I received a phone call from Côte d'Ivoire. I hate phone calls from

Côte d'Ivoire, because they normally mean bad news. Ilka, who was directing our project in the forest, was very worried. Over the previous three days, coughing and runny noses had spread to all the chimpanzees, and some were starting to show signs of serious weakness, to the point where they were unable to climb trees. After a short discussion, we agreed that I would fly to Taï as quickly as possible, accompanied by a veterinarian who had been working with our project for some time. In the meantime, the field team would put in place all emergency measures to protect the chimpanzees and document the progress of the disease: they would keep a twenty-four-hour watch on the weakest individuals sleeping on the ground at night to prevent them from being eaten by leopards, all seriously ill individuals would be followed for as many days as possible so that we could track the evolution of the disease, all field clothes and boots had to be disinfected each time anyone entered or left the forest, a safe seven meters' distance had to be kept from the chimpanzees to prevent any possible further disease transmission from us, and no one without professional qualifications was to touch any chimpanzee if one happened to die.

Sitting on the plane the following night, I recalled my anguish in the same situation five years before, when the dreadful Ebola virus had hit our study group. In all, twenty chimpanzees had died! We were caught totally unprepared, and we did not collect the data that would have allowed us to understand what was happening. I vividly remember the acute sadness and helplessness in the voice and eyes of my oldest and most experienced African observer, Grégoire Nohon, as he told me of the death of Ondine, one of our most dominant females. "I was totally crushed to see Ondine lying dead on the ground. There were in fact many chimpanzees around her, and I sensed the same sadness in their attitude, but I was blinded by my own feelings, and I cannot tell you anything more about what was happening. I am sorry, but I was too sad to be able to watch that scene any longer and I had to leave." The scientist in me was furious, because when they went back to the forest the next morning, they found Ondine's body entirely covered with leafy

DISEASE IN WILD CHIMPANZEES

As our close relatives, chimpanzees are vulnerable to human diseases. Are those diseases that are transmitted from humans to chimpanzees, as well as those that occur naturally among them, a real threat to their survival or not? We have been very slow to answer this question, mainly because we need to perform autopsies on dead chimpanzees, which are extremely difficult to find in the forest. When we find them, moreover, they are normally in such a state of decomposition that it is almost impossible to establish the causes of death. To overcome this problem, ever since the disease outbreak reported in this chapter, we have had a trained veterinarian present in our research camp who can collect samples as soon as possible after a chimpanzee dies. We have found that along with predation by leopards and poaching, infectious disease is one of the main causes of mortality in chimpanzees.

During the Ebola outbreaks in 1992 and 1994, 20 out of 52 individuals in one community of the Taï chimpanzees disappeared and were presumed dead. In two respiratory outbreaks (1999 and 2004), 14 out of 76 individuals died, and anthrax has been the cause of death for one to three individuals per year. A long-term analysis of demographic patterns (births and deaths) of the Gombe chimpanzees in Tanzania also came to the conclusion that disease was responsible for 58 percent of the deaths with known causes during a 47-year period.

Some of the agents responsible for chimpanzee mortality are pathogens indigenous to the forest. Often these are totally new to science, which, while interesting, makes our efforts to understand the cause of death in individual chimpanzees very difficult. For example, we discovered that a totally new strain of *Bacillus anthracis*, which causes anthrax, was responsible for the death of the Taï chimpanzees. We ▸

▸ also discovered a new strain of *Herpes* virus and a new type of *Streptococcus* bacteria. Such diseases are thus a natural component of their lives in the forest. However, of most concern, we also found that some of the pathogens were of a human origin. For example, two viruses responsible for flu in humans were found in dead Taï chimpanzees. Without a doubt, it is our responsibility as scientists to minimize the risk of disease transmission from humans to chimpanzees. Similarly, all ecotourism projects involving habituated apes must adopt proper hygienic measures to reduce the risk of disease transmission.

Ebola hemorrhagic fever (EHF), which has been killing high proportions of gorilla and chimpanzee populations in Central Africa, requires special discussion. Since the late 1970s, we have known that the Ebola virus is very deadly to humans, and so far no cure for it exists. We have no idea where this virus is hiding, how it is transmitted to us, or what causes the outbreaks, but we now have found that great ape populations have fallen victim to it in large numbers. In some remote regions of Central Africa where apes are numerous, up to 95 percent of them may have been killed by Ebola. It is estimated that between 1983 and 2000, the number of gorillas and chimpanzees in Gabon was reduced by more than half as a result of the combined effects of Ebola and illegal hunting. Sadly, Ebola is equally active in much of the neighboring Republic of the Congo, where the largest remaining population of great apes in Odzala National Park was ravaged by it in 2005.

Although diseases have always been a natural cause of mortality in wild great apes, given the tremendous increase of human populations throughout Africa and the increase of hunting, disease may now contribute dramatically to their decline and extinction if we do not act swiftly to protect them.

branches! Who had placed those leaves on her body? Was it Brutus, Ondine's long-time male supporter? And why did the chimpanzees do that? But, at the same time, the man in me understood; I would also have been crushed by the sight of her dead body.

Following this event, I was lucky to find Pierre, a veterinarian working in Abidjan, who was willing to go out of his way to help us understand what was happening to the chimpanzees. We started a joint project to study how the chimpanzees got infected and died, and how different diseases affect them. This collaboration also opened our eyes, because as trained biologists or conservationists, we often do not think in terms of disease when we think about the mortality of wild animals. For us, natural death can be attributed to predation or old age, and unnatural death to human hunting or habitat destruction. However, recurrent outbreaks of disease in Taï chimpanzees taught us that we were overlooking a very important dimension of the mortality in wild animals. Could it be that we had infected them with one of our viruses? This thought bothered me for many months, and I decided to find out what was going on. Thanks to our long-term studies of habituated apes, we are in a position to document the causes of natural mortality, and we now realize that disease kills a much larger proportion of them than previously suspected. If we are to help them, we need to understand what the main causes of mortality are in wild chimpanzees.

After a tiring flight, I landed in Abidjan, where Pierre was waiting for me, at 2:00 A.M. After four hours of sleep, we started the ten-hour drive to Taï forest. At dusk, we arrived at the camp and immediately sat down with Ilka to talk about the situation. The first shock came quickly; Loukoum had died the previous day of the respiratory disease. Furthermore, Lefkas was in very bad shape. He was unable to climb a tree and was sleeping alone on the ground. At that moment, two field assistants arrived from the forest and informed us that two additional bodies had been found; one probably was Castor, a female with a very handicapped leg, and the other was possibly an adult male.

Knowing that some of the runny nasal fluid was whitish yellow, a clear sign of a bacterial infection, we decided to try to dart Lefkas with an antibiotic before it was too late. An hour after arriving at the camp, we walked into the pitch-black forest that I knew so well from having walked through it for years with the chimpanzees. We now had walkie-talkies and could inform Grégoire and Nicaise, who were standing guard near Lefkas, that we were coming. We saw that they had hung a tarp as a rough tent some thirty meters away from Lefkas and placed a lantern fifteen meters from him, in the hope that it would keep any leopard at bay. Lefkas was in a pitiful condition, lying on his belly, breathing noisily through his open mouth and barely moving as we approached him. We were immediately afraid that we had come too late. Pierre rapidly succeeded in giving him an antibiotic shot via dart, and we simply had to hope that it would still have an effect.

I spent that night alone with Lefkas, the longest and saddest night of my life. He was obviously having great difficulties breathing, he could not sleep, and he changed position regularly. I looked at him from time to time, but let him be as quiet as possible. I saw the cut on his left ear: he lost a piece of his ear when he was a baby on his mother's belly as she fought to get meat from the others.

Sitting in the dark, I thought about the development of the chimpanzee project. When I started the project in 1979 with Hedwige, my wife, we were aware that the future of the chimpanzees was very uncertain. In the first years, we would occasionally hear bulldozers of logging companies illegally entering the park to extract trees, some of which were important food sources for the chimpanzees. Thanks to the relentless support of some of the Ivorian authorities, it was possible to stop these loggers, but gunshots were still heard from time to time. Poaching has always been a big problem, and habituating animals to humans could increase the risk of them becoming targets of poachers. How responsible was it of us to start this project? At the same time, chimpanzees had never been studied in a rain forest before, and their lifestyles in this environment were totally unknown. Over the years, we were able to show that the forest chimpanzees in Taï make and use more tools and hunt more cooperatively than chimpanzees living in more open woodland habitat, such as those Jane Goodall studied so well in Gombe National Park, Tanzania. Our discoveries shed new light on the theory of human evolution, which at the time was believed to be a result of adaptation to savanna. Our research was also instrumental in showing that chimpanzees exhibit different behavior in different environments. Furthermore, some of these behavioral differences are not explainable by ecological differences and are therefore a clear hint of the existence in this species of culture, something that had been thought to be a

CHIMPANZEE LIFE COURSES

Chimpanzees have life history patterns that are in many ways very similar to those of humans. Females are pregnant for eight months and mostly give birth to one infant at a time. Twins have been observed, but are relatively rare. Infants feed on breast milk for up to five years, during which time the mother carries the baby first ventrally (holding against her stomach) and then on her back. Only after five years, when the mother is pregnant with another infant, will she wean the current one. Once weaned, the infant is considered a juvenile, but still spends all its time near the mother for an additional five years. Only at ten years of age do young chimpanzees start to spend time away from their mothers. Females have their first baby when they are about thirteen years old, whereas males can be considered adult when they are fifteen. An adult female chimpanzee is typically accompanied by three of her offspring (nursing infant, juvenile, and young adolescent), which also provides the opportunity for juveniles to play with and handle their baby siblings.

Male chimpanzees remain in the community in which they were born (natal group) all their lives and continue to have regular contact with their mothers. This relationship between mother and adult sons lasts for the entire life of the mother, who can be an important ally for her ▸

uniquely human phenomenon. We hoped that by making such new observations on forest chimpanzees available to as many people as possible, we could contribute to securing a future for these fascinating animals, our closest living relatives. Different film crews, including David Attenborough's from the BBC, helped us make the Taï chimpanzees popular. However, it is very difficult to measure such benefits versus the costs we were imposing on the chimpanzees, and we were never sure if we had made the right decision. Now having to sit in the dark with a youngster I had known since he was a tiny baby, I was caught

▸ sons as they try to climb the male dominance hierarchy. On the other hand, females tend to migrate out of their natal group when they are between ten and twelve years of age and join a new group before they start to reproduce. This is generally viewed as a natural mechanism to avoid incestuous matings between fathers and daughters, as well as between brothers and sisters. However, when looking across different populations of chimpanzees, we see different patterns of female emigration. Among Gombe chimpanzees, about 50 percent of the females remain in their natal group, while almost all females transfer to a new group in the Taï and Mahale populations. The reason for this difference remains a mystery.

Determining the maximum life span of chimpanzees requires long-term observation, and since they are a long-lived species and females transfer between groups, this remains an open question. The earliest research on chimpanzees began in Gombe National Park, Tanzania, in the early 1960s, while Taï chimpanzees have been studied since the late 1970s. Currently we think that chimpanzees in the wild can live over forty years, with the oldest individuals reaching fifty. In captivity, chimpanzees have been known to live to sixty. In some cases, females become less fertile once they are over forty, but the majority of them still have infants at a very advanced age.

by surprise. Were we, the researchers committed to them, responsible for the disease and thus actually ignorant killers? Could we help them in any way? Would we be able to understand how such diseases were transmitted to them?

Night life in the forest is a very special experience, with all the noises, such as the large hammerhead fruit bats making their courtship calls, the little bushbabies with their swift sewing-machine calls, the tree hyraxes with their rhythmic crescendo calls, or the soft duetting first of the (partridge-like) francolins, and later of the (cuckoo-like) coucals, and the loud laughter-like calls of the Hagedash ibis flying over the dark forest. However, with Lefkas's laborious breathing near me, the experience of this night was different. Lefkas had stopped breathing by 5:00 A.M., and in my heart, a piece of the magic of the Taï forest vanished. As the daylight entered the forest an hour later, I realized that a little chimpanzee had descended from a nearby tree. It was Léonardo, Lefkas's tiny two-year-old brother. Seemingly totally stunned by this second death within two days, he glanced at his dead brother and then, not knowing where the other chimpanzees were, headed north alone. I did not expect to ever see him again.

LIFE GOES ON

Léonardo would not survive. Remarkably though, five days later, I saw Mognié carrying him on her back. She was only nine years old, however, and she could not breast-feed him, which was what he desperately needed now.

Genetic analysis later showed that Mognié's younger brother, Mozart, was actually Léonardo's half brother; they had the same father, Macho. Probably Mognié did not know that Léonardo was the paternal half sibling of her younger brother. Were Lefkas and Mognié such enthusiastic playmates because they were the same age, or because they felt they were somehow related? Did Mognié care for Léonardo because she knew he was her brother's sibling or for another reason? We'll never know what Mognié knew about her family relations.

Mognié immigrated into an unknown group two years later, which is very typical for a female of her age. She most likely joined the large neighboring group to the east, which is deeper inside the park and contains many large adult males. Female chimpanzees seem to transfer into groups containing many males, where they can find many good mates and can count on strong male support to assure a healthy social life for their progeny. Mognié was now living under the best conditions to start her own dynasty and have her sons become successful males as she became a high-ranking female.

Eight years after the deaths of Loukoum and Lefkas, and after much painstaking laboratory work, we were stunned to find a human virus in the samples taken from their bodies. People had infected them. All my old fears were confirmed, although it was impossible to be sure who was responsible—were we researchers to blame, or had it been poachers traversing the apes' home range? On further consideration, however, we realized that although we might possibly have infected them with human viruses, our constant presence had simultaneously been protecting, not only the chimpanzees, but all the other animals in the forest from poachers. Although nothing would bring Lefkas or Léonardo back to life, the densities of all large mammals were much higher at our study site than in other regions of the park. Luckily, the Taï forest is a relatively large, well-protected national park, and as long as poaching is limited, chimpanzees like Mognié will still flourish deep inside it.

FURTHER READING

Boesch, C. 2008. Why do chimpanzees die in the forest? The challenges of understanding and controlling for wild ape health. *American Journal of Primatology* 70:722–26.

Köndgen, S., H. Kühl, P. Goran, P. Walsh, S. Schenk, N. Ernst, R. Blek, P. Formenty, K. Mätz-Rensing, B. Schwieger, S. Junglen, H. Ellerbrok, A. Nitsche, T. Briese, W. Lipkin, G. Pauli, C. Boesch, and F. Leendertz. 2008. Pandemic human viruses cause decline in endangered great apes. *Current Biology* 18:1–5.

Leendertz, F. H., G. Pauli, K. Mätz-Rensing, W. Boardman, C. Nunn, H. Ellerbrok, S. A. Jensen, S. Junglen, and C. Boesch. 2006. Pathogens as drivers of population declines: The importance of systematic monitoring in great apes and other threatened mammals. *Biological Conservation* 131:325–37.

Walsh, P. D., K. A. Abernethy, M. Bermejo, R. Beyers, P. De Wachter, M. E. Akou, B. Huijbregts, D. I. Mambounga, A. K. Toham, A. M. Kilbourn, S. A. Lahm, S. Latour, F. Maisels, C. Mbina, Y. Mihindou, S. N. Obiang, E. N. Effa, M. P. Starkey, P. Telfer, M. Thibault, C. E. G. Tutin, L. J. T. White, and D. S. Wilkie. 2003. Catastrophic ape decline in western equatorial Africa. *Nature* 422:611–14.

Williams, J. M., E. V. Lonsdorf, M. L. Wilson, J. Schumacher-Stankey, J. Goodall, and A. E. Pusey. 2008. Causes of death in the Kasekela chimpanzees of Gombe National Park, Tanzania. *American Journal of Primatology* 70:766–77.

THREE

Encounters with Bili Chimpanzees in the Undisturbed Gangu Forest

CLEVE HICKS

June 29, 2005

It is morning in the Gangu Forest of the northern Democratic Republic of the Congo (DRC), and the drumming of the local chimpanzees resounds off the trees along the swampy forest clearing by which we are camped. The apes habitually announce their presence to one another in this way by rhythmically pounding on the buttress roots of trees. The familiar cacophony arouses me from my slumber and sends me lurching into preparation for the day's work. I feel my heartbeat speed up as if to keep pace with the drumming, and I struggle with my clothes inside the cramped, stuffy tent. In terms of getting me on my feet and out into the rain-soaked forest, the drumming is much better than a mug of coffee, which we have run out of anyway. Maybe, just maybe, we might encounter the makers of this primal music today. Out here, forty-five kilometers from the nearest road or village, the apes are free to indulge in their exuberant sound-shows with little fear of interruption by shotgun-wielding humans. The open savanna is far behind us now. We are pushing into the farthest reaches of this oasis of deep, green Unknown, deeper than I had imagined would be possible. We have come so close to the end of our transect work . . . but the rains are increasing, the floodwaters are rising, and we are soon going to run out of rice and beans. We cannot afford any more delays. In just two weeks the small missionary plane will be arriving to spirit me back to Holland with my precious data and films. Much as I would love to slosh east across the herb-choked *bai* and contact this morning's forest percussionists, our priority now is to finish the transect work.

• • •

ESTIMATING APE POPULATION DENSITY USING LINE TRANSECTS

How many apes live in a particular area? For a variety of reasons, this seemingly simple question is one of the most difficult ones for conservationists and scientists to answer. In order to estimate the population size of these animals, a researcher can extrapolate from the number of nests encountered on a transect to the probable number of nests in the forest area as a whole. A transect is a line cut along a pre-planned route through the forest, chosen at random. A transect can be cut using a GPS to ensure the maintenance of a straight line, parallel to other transects and separated by a suitable distance. After one team cuts the transect, a second team walks along it to count the nests.

Adult great apes generally construct a fresh nest every night in which to sleep. By measuring the perpendicular distances of nests or groups of nests from a transect, it is possible to calculate the effective distance of nest detection, which is the distance beyond which the number of observations made equals the number of observations missed. Obtaining the effective distance allows one to calculate the density of weaned individual apes in the survey area, provided that one has information about certain parameters: the number of nests each ape constructs on average per day, and the amount of time it takes for nests to disappear (the decay rate). The calculation can be done using the ▸

I first came to this area a year ago, with the goal of uncovering the secrets of the chimpanzees of the northern DRC. Chimpanzees have been studied in many areas across eastern and western Africa, but here in the middle of the continent, over 600 kilometers from the nearest long-term chimpanzee study site, virtually nothing was known about our forest cousins. When I arrived, I had several questions that I needed to answer. In the first place, were there any chimpanzees at all here,

▸ Distance formula, available online (http://en.wikipedia.org/wiki/Distance).

One drawback of cutting transects is that it effectively creates nice trails, facilitating poachers' access to the area. The last thing we wanted to do was open up this pristine forest to poachers. In order to minimize the possibility that elephant hunters might use our transects to gain easier access to the Gangu Forest, we deliberately left large stretches uncut in the savanna and marked them only with flagging tape, which we later removed on our return journey.

Transect work is difficult and time-consuming, particularly when it involves traveling into remote and inaccessible areas such as at Gangu. However, short of actually habituating the apes and counting them one by one, it is one of the few ways we have of estimating ape densities. It is crucial that we come up with accurate estimates in order to make the best allocation of scanty conservation funds. Currently, other methods of censusing great apes are under development, such as counting individuals recorded on film by researchers and/or trap cameras (motion-triggered cameras), and recording the number of ape vocalizations using audio equipment planted in the forest. For some areas that are impractical or impossible to visit, the probable sizes of ape populations can be estimated by analyzing different parameters of habitat suitability (forest type, human disturbance level, climate) using satellite images.

and if so, how many? Absent any surveys, several maps of chimpanzee distribution have left this area—hundreds of thousands of square kilometers of virtually unpeopled forest and savanna—a large blank: a tantalizing question mark. Given the enormous size and remoteness of this stretch of potential ape habitat, there was a high likelihood that we would be adding tens of thousands of chimpanzees to the official tally, providing a glimmer of hope, given that the species' population is dwindling toward extinction across most of Africa. It was even rumored at the time that I joined the project that the Bili chimpanzees might share their forest with a relict population of gorillas—the only gorillas in the central DRC—or that the two ape species might even hybridize here.

For this reason, my team and I are now cutting and walking transects, counting the night nests made by the apes, in order to estimate their population densities. We also want to know more about the eating habits of these apes, to confirm or refute the claim that they eat big cats, and to decipher their social structure. My personal passion is chimpanzee behavioral diversity. Recently, it has become widely—although not universally—accepted that like their human relatives, chimpanzees possess culture (i.e., populationwide socially transmitted behavior). I hope to document an all-new set of chimpanzee traditions in the region. Eventually, we plan to set up a long-term research project in the forests of Bili, following and observing generation after generation of great apes, as has been done in Tanzania, Côte d'Ivoire, and other countries.

• • •

My team and I pack our bags and prepare to follow the next stretch of transect. Although it is not the priority of our current work, I am in the mood for an encounter with some chimpanzees. We can only hope to run into another group of the apes further to the west, which is not unlikely considering the high density of their nests that we are finding in this virgin forest. For the eight months prior to the start of our transect work, while we were installed at Camp Louis and attempting to follow the chimpanzees rather than count them, the head tracker,

Ligada Faustin, captivated me with his descriptions of the human-free "enchanted" forest of Gangu, full of elephants, bongo antelope, and "naïve" chimpanzees that did not flee from humans. In years past, during long fishing jaunts there with his family, Ligada said, he had been approached on the ground by the resident chimpanzees, who would go so far as to investigate the contents of his backpack.

Even given the possibility that Ligada was embellishing his story a bit, this type of reaction by the chimpanzees would stand in marked contrast to the behavior of the chimpanzees that we tried to study during our first eight months at Camp Louis, within ten kilometers of Baday Village and a lightly traveled dirt road. Almost without exception, within seconds of spotting us, the adult males would fling themselves in panic from the trees to the ground below and run away. When they did not flee immediately, females and juveniles appeared to be frozen with terror in the trees, screaming, while a rain of diarrhea (a common indication that chimpanzees are frightened) showered down around us. Clearly, the chimpanzees of Camp Louis had had some unpleasant encounters with the local villagers. For this reason, we aborted our efforts to habituate the Camp Louis apes.

The DRC is an extremely unstable country. At the time of our survey, Bili had a low human population density with large undisturbed areas safe for wildlife, but we knew that this could change rapidly with the entry of rebel militias, illegal loggers, or (as would eventually be the case) gold and diamond miners. These intruders would quickly open up the region to the commercial bushmeat trade (*bushmeat* means meat from wild animals), which elsewhere is rapidly emptying the DRC of its wildlife. In this context, we thought it unethical to deprive these chimpanzees of their fear of humans by getting them used to our presence.

Instead, we decided to visit Ligada's "chimpanzee wonderland," Gangu, and meet its fearless apes (if his stories were true). At the same time I wanted to explore it as systematically as possible so that we could estimate the chimpanzee density. An advance team led by our field assistant Makassi would cut three parallel 55-kilometer straight-line transects from the road out into the untouched forest; as we walked along the transects, we would count ape nests, elephant dung, and signs of humans.

Having entered the eastern edge of this remote forest, we see that so far the Gangu chimpanzees' reactions have been different, just as Ligada promised. During our encounters with them, the adult males have usually stayed on the scene, with their hair bristling, peering down at us with guarded curiosity. Youngsters have even blithely plucked and eaten fruits as they watched us from fifteen meters away. This strange and wonderful phenomenon began to occur when we were about twenty kilometers from the road, and it has become more pronounced the further west we have walked. According to Ligada, it should get even better up ahead, and the chimpanzees might even approach us on the ground. He has rarely been wrong about anything like this before . . .

Not only the chimpanzees of Gangu seem oblivious to the dangers of humans. A few days ago, we inadvertently came between a group of grunting, snorting, sneezing red river hogs and the leopard who was stalking them. When we heard the leopard's rumbling growl approaching us from the south, Ligada and I crouched down, our cameras in hand, poised to catch the predator on film. But when the bushes shook and the massive spotted head rose out to peer at us from fifteen meters away (it undoubtedly mistook us for the hogs), we found ourselves frozen beneath the lordly gaze of the big cat. By the time we had emerged from our collective trance and raised our cameras, the leopard had melted back into the undergrowth. Ten minutes later we heard the hogs scream with panic not far to the north—apparently the leopard had circled around and achieved its goal.

Today, after a quick and perfunctory breakfast of beans and rice, we set out into the drippy, rain-soaked forest, knowing that it will be a wet day. We are roughly following the course of the Gangu River through increasingly swampy terrain. Luck was against us on our two northern transects—despite frequently hearing chimpanzees, we failed to contact them further than thirty-five kilometers west of the road. This trip is our last opportunity of the season; it will be at least a year before we can return. Ligada takes the lead, wielding only his machete, scanning the partially obscured skyline for the silhouettes of chimpanzee nests. He doesn't miss many. Old Chief Mbolibie follows him, then I and another spotter, and finally two teenage porters.

As we pick our way along the shoulder-width transect hacked out a few days before by Makassi's cutting team, I stop to count and measure the numerous piles of fresh elephant dung, the occasional buffalo patty, and the dozens of chimpanzee night nests, the latter often clustered together in what the assistants referred to as "villages" around low-lying streambeds. The chimpanzees here have a peculiar preference for constructing large, complex ground nests. Gorillas—but not chimpanzees—are well known to build many of their nests on the ground; in fact, this was what had led earlier explorers to posit the presence of gorillas in the area. For the Bili chimpanzees, it is a nearly inexplicable custom, given that elsewhere chimpanzees nest almost exclusively in the trees. It is especially puzzling considering that these forests are teeming, not only with leopards, elephants, and buffalo (none of which you would like to have stumble across you in your ground nest late at night!), but also with lions and hyenas, more typically savanna predators. A year from now, Ligada will be fortunate enough to observe a Gangu chimpanzee eating a leopard carcass, and will bring me back a paw as evidence: might this explain how the apes can get away with ground-nesting?

Whenever we encounter a nest, the spotters and I slowly pace out twenty meters ahead on the transect, looking for more nests. Then two of the trackers fan out on either side to search for any that we might have missed. Ligada and I measure the perpendicular distance of each nest from the transect, and then the circumference of the nest trees.

As we are by now running low on rations, we frequently stop and stuff ourselves full of delicious, minty, light green zingi fruit *(Parinari excelsa),* which at the moment are scattered across large strips of forest floor. Ensuring that we bring enough food is always the toughest part of planning these long-distance treks. I do not allow the assistants to hunt, and "dam-fishing" (one of the greatest delights of the Azande, the local people with whom we work) is only practical in the dry season; therefore, carrying rice and beans in bulk is our only option. But the further we go and the longer we stay out, the more food we need and the more porters we must hire to carry the food. The porters themselves eat a lot, so by the end of a transect, we are usually running low on vittles. The trackers supplement our diet by gathering succulent wild mushrooms and spinach-like kalembai swamp herbs, but after a while the monotonous repertoire of beans for breakfast followed by beans for dinner, and then beans for breakfast again begins to wear on our palates and our stomachs. As the trackers might say, it's tough—"Pénible!" This tedious diet frequently leads me to hallucinate about falafel and pasta, which is why we now are extremely grateful for the tart, sweet zingi fruit, as are the chimpanzees, hornbills, and monkeys.

At the end of our work day, as the sun descends behind the tree line, we arrive at the edge of the flooded Bilima River. According to the trackers, this waterway will soon rise by several meters and become filled with Nile crocodiles. Bili is highly seasonal: during the dry season, when for months not a drop of rain falls, many of the local streambeds dry up completely, and finding drinking water can be difficult. Now we have the opposite problem, too much water, slowing our travel. According to Ligada, the really naïve chimpanzees live on the other side of this natural riverine barrier. Even the fearsome hunters of Gumbu (a village about forty kilometers southeast), who occasionally venture into these forests during the dry season, rarely make it out there.

We set up camp in the middle of a wide and spacious elephant *route royale,* or "king's highway." Out here, the only paths we see were made by elephants: large spacious trails free of vines and undergrowth cleared by generations of passing feet and trunks. Elephant dung is everywhere, and based on the various sizes, we can see that it was deposited by mothers, babies, "big papas." . . . Perhaps it is a foolhardy place to sleep, but at least we don't have much undergrowth to clear.

Huddled next to the flickering fire of our river's-edge bivouac, I peer out across the swirling dark currents of the Bilima and imagine the untouched wilderness that awaits us on the other side. I am lulled into a trance by the murmuring bee-like buzz of the fruit bats foraging up in the canopy. A hyena adds its baleful whoop to the nighttime concerto. Later, in the early morning hours, we'll be awakened by an ape drum concert from the far side of the river. Ligada's "fearless chimpanzees" await us!

June 30, 2005

This morning we are obliged to slog 100 meters through the waist-deep waters of the Bilima, pushing our way against the powerful current and securing our footing using walking sticks cut from saplings. The swamp forest bordering the river has flooded from the incessant rain, swirling up around the exposed roots of the uapaca trees. To obtain more information about the nesting habits of the local chimpanzees, we are taking measurements on the trees supporting the dozens of chimpanzee nests that we encounter as we walk. We slip and slide over roots of the uapaca as we careen from nest to nest. It seems that every time we have triumphantly taken our final measurements on the last nest, Ligada's keen eyes spot yet another lurking further off in the canopy above the treacherous swamp. Although it is exciting to be finding such a large number of chimpanzee nests, the swamp is slowing us down, and we have only a day or two before we must return to the village. By the time we reach the far shore of the flooded swamp, having lifted our boots over innumerable floating clumps of elephant dung, it is nearly noon and we are exhausted. We are relieved to be on dry land again.

Once out of the river, we are excited to encounter several sites where one or more chimpanzees have recently used tools. Over the years I have developed a keen eye for spotting these: small stripped sticks projecting up out of ant holes. Often we are tipped off by clumps of leaves that have been plucked from the tools. A brief search will reveal the nearby saplings from which the tools were ripped. The ants at today's sites have disappeared, but by examining the structure of their homes, we can be certain of one thing: these were not colonies of *Dorylus wilverthii*. Those formidable, aggressive ants, which travel in millions-strong formations and consume every small animal in their path, are intimately familiar to us, thanks to their charming habit of flooding into our tents in the middle of the night and biting us—not a pleasant way to be woken up, especially when one then has to flee out of a tent door that is covered in a writhing mat of the ferocious insects. To catch that type of ant, the Bili chimpanzees fashion enormous tools measuring up to 2.5 meters in length and thrust them into the cavernous holes under the ants' massive red earthen mounds. When the ants stream up the stick, the chimpanzees are able to scoop them off at a safe distance and eat them. We still do not know why the Bili chimpanzees make their tools so long. At other study sites, tools for extracting aggressive driver ants of the same or related species as *Dorylus wilverthii* are almost never more than a meter long. For nonaggressive ants at Bili, the chimpanzees use tools only half as long on average. Such was the case at this site. We collect and measure these shorter tools and continue on our way, finding and measuring more chimpanzee nests. To the southwest, we hear periodic outbursts of chimpanzee drumming and pant-hoots. It would be tempting to go after them, but for now we must stick to our transect work.

By late afternoon we have arrived at another big muddy saline (a cleared open area of swamp), the Vugu, this one a bona fide elephant stomping ground! Dozens of road-like elephant paths criss-cross the saline; fibery clumps of their fresh dung are mixed with churned-up mud at the water's edge. The dung is warm: apparently we have just missed the animals as they emerged from the swamp and fanned out south into the forest. But no matter: an eruption of chimpanzee pant-hoots greets our ears from just a few hundred meters south. We are at tonight's camp site and it is a good time to end our transect work for the day—let's go meet some Gangu chimpanzees! Flinging off our back-packs, Ligada, Mbolibie, and I plunge into the dense forest skirting the edge of the saline, leaving the rest of our team to construct our bivouac and cook dinner. A chorus of screams and pant-hoots from the trees ahead makes us quicken our pace. We speed our progress by using the many elephant trails we encounter. The goal is to catch the apes before they descend to the ground and move off. Ligada leads, as always. He has the uncanny ability to spot chimpanzees long before I do, and then, with the finesse of an orchestra conductor, maneuver me into exactly the right position to film them. As we approach the apes we hear the pant-hoots and tree-drumming begin to recede into the distance. This time we have just missed them: beneath a zingi tree, we find the pale green pits of hundreds of the freshly devoured fruits. The chimpanzees have just enjoyed a huge feast, then moved on, most likely without hearing us coming. We know from experience that if they had been fleeing us, they would not have continued vocalizing to one another as they moved off.

Dejected, we sit with our chins on our knees and listen. Ligada and Mbolibie tell me that it is hopeless, and that we might as well return to camp, but I am stubborn, and we spend another thirty minutes listening with keen ears for any hint that the chimpanzees might still be in the area . . . but we are rewarded with only a few faint tree-drummings far to the south. In situations like this, I think of Bilbo Baggins and his dwarf friends in *The Hobbit,* in the forbidding forest of Mirkwood, after they unwisely allowed themselves to be lured off the trail, drawn forth by the merry sounds of the Wood Elves' feast. Each time they approach a glade full of carousing Wood Elves, the elves hear them coming, kick out the fire, collect their food and spirits, and take their party to a new location, leaving Bilbo and friends stumbling around in the dark. On a bad day that is exactly what following chimpanzees is—Mirkwood Forest meets the DRC. I sigh, agreeing with Ligada and Mbolibie that we ought to be heading back to camp before it gets dark, because we do not know this forest and do not want to get lost with elephants roaming around. Just as we begin to retrace our path to the north, a renewed and ebullient chorus of pant-hoots bursts out of the forest just fifty meters to our south. They were that close the whole time, I imagine, digesting quietly, their bellies full of zingi fruit, reclining on the limb of some mighty tree. Ligada, Mbolibie, and I exchange gleeful grins, spin around and race south through the forest. We won't let them slip away this time! I allow myself a teasing "told you so" to Ligada.

A wild commotion is developing just ahead: we hear a flurry of deep, guttural barks, followed by a long series of screams; it sounds as though several apes are thrashing about through the undergrowth. What could they be doing? We can almost see them behind a tangled curtain of vines and herbs. I have turned on my video camera, but I am filming only the wall of vegetation between ourselves and the apes. We listen spellbound as a large chimpanzee gallops across a clearing a few meters away, barking (almost roaring!) and beating down saplings, his knuckles slapping the soft soil as he runs. Another individual, the screaming one, who we suspect is the victim, races south. The chimpanzees are fighting over something, I am sure of it.

Once the sounds subside, we gingerly tiptoe our way forward, stepping high and light to avoid breaking sticks underfoot and giving away our presence. We emerge into the clearing just vacated by the apes (once again, we think they were embroiled in their own noisy drama and thus completely unaware of our presence) . . . and find the freshly devoured carcass of a tree pangolin *(Phataginus tricuspis)* lying at our feet. A pangolin is a tropical ant– and termite eater resembling a cross between an armadillo and a sloth: they have the look of large rubbery pine cones with prehensile tails. This animal's carapace has been cracked open, the head bitten off, and the flesh and innards scraped out with incisors. We think it likely that the chimpanzees ate this pangolin; the trackers point out that if a lion, leopard, or hyena had been responsible, it would have wolfed down the entire animal, scales and all (and indeed we had often encountered pangolin scales in hyena and leopard dung). In the soil a meter from the carcass, a chimpanzee has left its "signature" in the form of a perfect footprint, complete with divergent big toe. Pangolins are rare prey for chimpanzees, documented only at two other study sites in west and central Africa. We cannot rule out, of course, that the chimpanzees might have scavenged this prey from another predator.

I am yanked out of contemplating this amazing scene by another volley of pant-hoots directly to the south. Leaving Mbolibie to guard the pangolin carcass, Ligada and I speed off in that direction. Normally, I would go slowly and let Ligada guide me . . . but I have a hunch that something interesting is taking place just ahead. Ignoring Ligada's cautionary gestures for once, I race ahead, camera at the ready. Fifty meters along, I emerge onto a big, clear elephant path. As I peer around, I think I hear a little cough or grunt very close to the south. Ligada catches up to me; he says that that the chimpanzees must have headed south, and that we should return to camp to be sure that we arrive before dark. I point in the direction from where I heard the cough, and then pick my way there cautiously through the foliage. More sounds. A sharp musty smell. Saplings rustle up ahead, and I see a black blur behind the green web of vegetation.

I immediately crouch and aim my camera in the direction of the movement. Through the lens I can see a grizzled gray adult male chimpanzee hunched over the ground only twenty meters away, poking something about in a hole, wearing an intense frown of concentration on his dark face. He is dipping for ants! Sitting directly in front of him and peering down with curiosity at the elder chimpanzee's actions is a pale, jug-eared youngster. The adult is bringing the ants to his mouth as if with a single chopstick, licking them off with obvious relish and then chewing rapidly. He repositions himself. The youngster leans a little too close, causing the adult to emit a gruff cough and then cuff the little upstart; it screams and races off to the south. This mini-drama has made the chimpanzees shift positions, and behind the male, I can now see an adult female with a baby clinging to her belly, also dipping for ants. I film the male returning to the hole and adjusting his ant wand. Alas, after about a minute the adults notice me and rush off quickly . . . even for the Gangu chimpanzees, a close-range encounter with a weird-looking bespectacled biped on the ground may be too much.

After waiting for a few minutes, Ligada and I move forward and investigate the tool site, collecting three short stick tools. The tools were plucked from nearby saplings and stripped of their leaves. They are covered in driver ants, but I am pleasantly surprised to find that I can handle them without jumping up and down and yelping in pain. These ants, *Dorylus kohlii,* are far less aggressive than their bellicose cousins *Dorylus wilverthii;* one might even describe them as wimpy! Their swarms are also relatively small. This is possibly the same species of ant for which we found the dipping evidence the day before.

I am pretty excited: we now have definitive evidence (on film!) that the Gangu chimpanzees use the "mouth-off" technique to harvest these less aggressive ants with short tools (this technique has been seen at other study sites such as the Taï forest and Bossou). This evidence will help us better understand chimpanzees' choices in making tools to hunt for various species of insects. We can predict that for the more aggressive species, they probably use the alternate "sweep through" technique (where the chimpanzee holds the tool in one hand and sweeps the ants off with the other), as has been seen elsewhere for long tools. Chimpanzees are quite capable of inventing tool sets to enable them to successfully exploit insect species with different defenses and differently constructed nests (see chapter 6). Perhaps they are using longer tools for the more aggressive species here to keep their distance from the swarming nest and avoid painful bites. Or perhaps they need the longer tools to exploit the deeper subterranean chambers of *Dorylus wilverthii.*

As we are reflecting on our tool discovery, Ligada taps my shoulder and points above my head. I look straight up into a pair of bewildered, hazel-colored eyes scrutinizing me from just a few meters up a tree. They belong to a juvenile chimpanzee female, probably the same youngster I just filmed getting scolded by the adult male. She has huge ears, a white face, and a tail tuft. She dangles for several minutes just out of my reach, occasionally inserting her index finger into her gaping mouth as she tries to figure out just what in the Gangu I could be. Perched nearby on a branch is an older, brown-faced youngster displaying similar insouciance. If these juveniles are showing any emotion at all other than curiosity, they seem a bit cross at us for disturbing their lovely "ant-scapade," and they briefly exchange a volley of wraah-barks.

I was never this close to chimpanzees at Camp Louis. The encounter lasts over half an hour, as the little female clambers from one perch to another trying to get a better vantage point from which to peer down at us. She finally gets a bit nervous when Ligada circles around behind me with his camera. She screams, more out of indignation than fear, before finally moving off with her companion. We then hear an adult pant-hoot from not fifty meters away; others in her group have remained within earshot.

Completely bowled over by what we have just seen, but also conscious of the rapidly failing light, we race back to camp using elephant trails. I collect the body of the pangolin, which we smoke over the campfire in order to preserve it. Dinner: alas, more rice and beans, but I am too elated to care! What a stroke of luck to witness so much chimpanzee behavior over so few days on my last week of this season in the Gangu Forest!

July 1, 2005

It is just before noon on our last day of transect work at Gangu and we are within a kilometer of the west end of our transect. Our luck is still with us. Ligada, Mbolibie, and I are standing beneath a treeful of boisterous, noisy Gangu chimpanzees. We walked right under them thirty minutes before, even observing a freshly spat-out zingi fruit at our feet, only to finally be alerted to their presence by a loud volley of pant-hoots while measuring a nest along the transect to the west. There are at least eight chimpanzees in the trees above us: several adult females, two adult males, a couple of juveniles, and what Mbolibie describes as "an old man." They fail to notice our arrival at first, and I film some thrilling images of a hulking black adult male bounding acrobatically across the canopy, following a desirable, pink-bottomed estrous female. When the chimpanzees become aware of us, they do not flee, but stop and stare. They then reposition themselves to get a better look. The last time I saw anything like this was in the Ndoki Forest in the Republic of the Congo, another remote undisturbed paradise (see chapter 6). But this is

Gangu. What looks like an old granny chimpanzee squints perplexedly at us; a youngster peeks over her shoulder. Another of the youngsters is peering down at us with an expression of sheer wonder, almost comically stupefied by the sheer novelty of it all! (My expression is probably more or less the same.) The juvenile then knuckle-walks tentatively toward us across a branch for a better look. Above us, a peculiar-looking adult male sporting a thick coat of shiny black hair everywhere except atop his pale bald head, positions himself in the crook of a tree and gazes down implacably upon us, projecting the air of an indolent king. "How dare you trespass in MY domain?" I can almost hear him say. He seems to be studying us, as we are him, and remains for a time even after his fellows have departed.

Later, while watching the film of the contact, I notice that this bald male appears to sport a sagittal crest down the center of his stout heavily muscled skull. We have seen this gorilla-like ridge of bone once before, on a Bili ape skull acquired by Karl Ammann. Professor Colin Groves, who has made systematic measurements of Bili ape skulls, claims that they are distinct enough in morphology to make them a different subspecies of chimpanzee. Geneticists are not convinced, however, and we shall see. This fellow certainly looks different. Following our long period of mutual assessment, the striking Gangu male appears to lose interest in us and slowly descends to follow his group. But they don't move far, because only an hour later, we hear them pant-hooting noisily just a few hundred meters away. This can be contrasted with the behavior of the Camp Louis chimpanzees—often, following one of those traumatic contacts, we would not hear a peep out of them for several days.

Returning to our companions Likongo and Garavura, whom we had left behind at the nest site, we find that they have fled several hundred meters back on the transect: from lions, they tell us. What a forest! It is going to be hard to leave it. But there is no time to waste. We have a long trek to the main road, and after that, I have a farewell party to throw and a plane to catch. An hour later, we reach Makassi's last piece of flagging tape, marking the end of our last transect. I hug my co-workers and congratulate them on our achievement. My biggest praise goes to Ligada. He is beaming with pride. Finally, on the last of the transects, on our last week in the forest, we found his fearless Gangu chimpanzees, and they were as spectacular as he said they would be!

• • •

Five sobering years later, I was determined to return to this "magic forest" as soon as I could. Chimpanzees free from fear of humans have become a vanishingly rare phenomenon in Africa. Studying them provides us with a fascinating glimpse of what the lives and cultures of these apes were like before *Homo sapiens* began carving up their habitats into fragments and shooting them for bushmeat. Surveys in other forests around the region showed that the northern DRC ape population is healthy and widespread. Excitingly, the set of behaviors shown by the Bili-Gangu apes (ground-nesting, ant-dipping, termite-mound and snail-smashing, and the use of leaf cushions) is remarkably uniform across an enormous area, providing more evidence that the apes are still connected in a large continuous population. We have been presented with a golden opportunity. I considered it crucial to establish a long-term research base in the area in order to keep the eyes of the world focused on Bili-Gangu. The Gangu Forest is officially gazetted as a game reserve, but it is protected on paper only, and has until recently been spared from the predations of humans only by its remoteness. Those days will soon come to an end.

A year after the transect project had ended, I returned to this untouched forest, where I established Camp Gangu in the center of the apes' homeland. I spent more time with the chimpanzees, and uncovered more of their secrets. Unfortunately, as we worked to open the forest to research and conservation, other interests were pushing equally hard to open it to exploitation. In 2007, we witnessed an invasion of thousands of illegal gold miners into the Bili-Uéré Game Reserve. The community conservation project that had been active in the area for years promptly shut down, along with our research project. Dejected but determined, I next spent a year and a half surveying the Buta-Aketi Forests about 200 kilometers south, where gold and diamond mines are proliferating in the forest and the market in chimpanzee meat and orphans is taking a terrifying toll on the regional ape population.

The animals of Gangu, apart from the more cosmopolitan, wide-ranging elephants, are familiar only with the local Azande people, who enter the forest in small numbers, focus more on fishing the streams than hunting, and appear to have a light ecological footprint. The miners, however (accompanied as they generally are by commercial bushmeat hunters) are radically different, and if they are able to establish large camps deep inside the forest, the Gangu chimpanzees and other fauna will be "sitting ducks." It is up to us to find a way to protect Gangu before it is too late.

FURTHER READING

Hicks, T.H. 2007. Into the world of the Bili apes: Bili field season 2004–2005. www.wasmoethwildlife.org/folder2004–2005/ (accessed July 21, 2010).

———. 2007. A new beginning: Bili field season 2006–2007. www.wasmoethwildlife.org/folder2006–2007/

———. 2008. Trading chimpanzees for baubles: A bushmeat crisis in the northern DR Congo. 3 vols. www.wasmoethwildlife.org/folder2007–2008/part1/index.html

Hicks, T.H., L. Darby, J. Hart, J. Swinkels, N. January, and S. Menken. In press. Trade in orphans and bushmeat threatens one of the Democratic Republic of the Congo's most important populations of eastern chimpanzees, *Pan troglodytes schweinfurthii*. *African Primates.*

Kuehl, H., F. Maisels, M. Ancrenaz, and E.A. Williamson. 2008. Best practice guidelines to surveys and monitoring of great ape populations. Gland, Switzerland: IUCN SSC Primate Specialist Group (PSG).

Morgan, D., C. Sanz, J.R. Onononga, and S. Strindberg, S. 2006. Ape abundance and habitat use in the Goualougo Triangle, Republic of Congo. *International Journal of Primatology* 27: 147–79.

FOUR

Is Blood Thicker Than Water?

GOTTFRIED HOHMANN AND BARBARA FRUTH

The call of the ibis moves through the canopy like a ghost. It's a sound typical of the forest at dusk, when sunlight disappears behind the wall of dense vegetation. Hearing it during the day makes us wary. Luckily, none of our field assistants are around. Mongo people hate this bird; they hate any animal that moves about when the forest gets dark. A nocturnal lifestyle is considered to be evil. The light of the forest is a dark green. The air is cool and dry and smells like mothballs. Nothing moves—almost nothing. Bonobos can melt into the forest. Dark fur disappears in the shadow of ground cover. The only visible landmark is the young adult male, Tagore, with a pitch-black face and shiny eyes. The ibis calls again, closer now. It sounds as if someone is being strangled. One of the female bonobos shifts in her day nest. The male looks briefly in her direction, scratches his long arms and moves his body to a comfortable position. He looks relaxed, but one can be sure that he is scanning for signals from the others with him.

Living in a society where individuals may choose with whom they spend their time can be advantageous, because it helps to avoid tension and conflicts. On the other hand, it creates gaps in an individual's knowledge of social relations between others that can have far-reaching consequences. Individuals may not see each other for weeks or months, and when they meet again, things may be different. Alliances and friendships, important because they can diminish the power of others, change with time. The bonobo social network is constantly in motion, but it contains kinship knots, which are resistant to change and may hold for a lifetime.

Male bonobos show a trait that is rare among most nonhuman primates and many other social mammals: male philopatry. This means that males spend their lifetimes in the community in which they were born, known as their "natal community." As a consequence, adult males are likely to have access to close relatives, or kin. Kinship is an important parameter and determines social relations in a predictable way: relatives are more likely to support each other and to cooperate than nonrelatives. Countless studies across the animal kingdom have found support for the predicted relationship between kinship and behavior. However, there are species that do not fit the theory, and bonobos are one of them.

Bonobos are closely related to chimpanzees. These sister species only diverged from a common ancestor a million years ago. Given their common evolutionary history, it is not surprising that the two *Pan* species share many traits. In fact, their anatomy and morphology are so similar that for a long time scientists considered them a single species. Similarities also extend to patterns of sociality and grouping: in both species, males are philopatric, while adolescent females leave their natal community and transfer to other communities. This means that within a community, females are less closely related than are males. Following the predictions of sociobiological theory, it would be reasonable to assume that close kinship ties lead to strong associations, affiliation, and cooperation among males, but not among females. Wild chimpanzees seem to follow the predicted path: males are attracted to each other and develop strong ties, the center of the chimpanzee universe. Males greet, kiss, and hug each other. Decisions are often made by male alliances, rather than by individuals. Communal defense, hunting, and warfare are important activities executed by bands of males.

SOCIAL RELATIONSHIPS BETWEEN FEMALES AND MALES

In contrast to most other mammals, adult males and adult females nearly always live together in social groups in most primate species. Among the great apes, we see remarkable diversity in the types of grouping patterns and social relationships. This leads to several questions. Why do they live in social groups? What are the costs and benefits of these permanent associations? Why do we see such variety in behavioral patterns among different species?

Animals often group together as a strategy to avoid predation. The more eyes and ears in one location, the greater the ability to detect predators. Predation is a danger even for the great apes; several cases of leopards killing chimpanzees and gorillas have been documented at different field sites. Groups of animals are better able to defend food resources or territories than a single individual, but this also leads to competition among group members. As explained in chapter 8, the risk of infanticide also is believed to be one of the factors responsible for male and female primates forming permanent associations with one another. In brief, males tend to kill unweaned offspring sired by other males, thus increasing their own reproductive potential. Not only does this reduce the reproductive success of competitor males, it also reduces the amount of time before a female is able to produce another infant. Since both males and females want to ensure the survival of their offspring, it is beneficial for the males to provide protection for their own mates and infants, which is best done by forming social groups.

A consequence of this risk of infanticide is that females prefer males who are physically and behaviorally capable of preventing invasions by other males. Primate males are characteristically larger in body size than females, which typically leads to a general pattern of male dominance over females. Silverback gorillas are always dominant over adult females, and chimpanzee males are dominant over females most of the time. However, bonobo females are generally dominant over ▸

▸ the males, even though bonobos and chimpanzees exhibit the same degree of sexual dimorphism (how much larger males are than females). Recent studies nonetheless indicate that social relations between the sexes can be complex and flexible. Male dominance, which determines access to resources, can be expressed in different forms, ranging from despotic to tolerant. In some species, such as the bonobo, males are reluctant to use their physical superiority in conflicts with females.

Yet without a doubt living together in a social group is not only about dominance and aggression. In most species of primates, females and males develop strong and lasting social ties, typically spending a large amount of time in close spatial proximity, supporting one another in aggressive conflicts, and engaging in high rates of friendly, intimate interactions such as grooming. Given that this bonding enhances individual males' mating opportunities, it may be regarded as a reproductive strategy. While infanticide and other threats to females remain strong selective forces for male dominance, some species have evolved alliances among females that are strong enough to reduce the effectiveness of male aggression and modify the degree to which males are able to monopolize or limit access to resources.

What are the ecological and social conditions that promote female dominance or co-dominance in primates? Concealing when they are sexually receptive may be the most efficient way for females to diminish the capability of males to control (and dominate) them. If males are unable to detect ovulation, their reproductive success depends largely on behavioral cues by females, which in turn means that females may be able to exercise more choice in mates. As a result, males who are tolerant of and invest in good relations with females are more likely to become mating partners than males who dominate females by using physical power. As reliable signals of advertising sexual receptivity (and estrus, the likely time of conceiving) disappear, selection favors males ▸

▸ that are tolerant and cooperative. Social relations between females and males shift from short-term consortships enforced by males to mutually-arrived-at social bonds between males and females. Such patterns of female dominance or co-dominance are not often observed in primates. Bonobos are one of the species in which they do occur, which is fascinating, given that they are otherwise so similar to chimpanzees.

Another factor that influences the strength of social relationships within and between the sexes is kinship. To prevent inbreeding, adolescent individuals of one sex, often males, transfer to neighboring groups. Consequently, interactions among adult kin are usually restricted to individuals of the same sex, often females. In species such as baboons and macaques, large matrilines of related females form the core of social groups. However, in species such as bonobos and chimpanzees, where males usually do not emigrate, males may live together with their mothers until well into adulthood. Evolutionary theory predicts that individuals will maximize their reproductive fitness by engaging in alliances with close kin, and the close bonds between adult males and their mothers that have been observed in bonobos and chimpanzees are in line with this theory. The biological significance of kinship for the success of adolescent and adult apes needs to be further explored.

Overall, social relations between the sexes in apes reflect a compromise between male and female mating strategies. Where these strategies meet along the scale from "demonic" males dominating the social life of females to tolerant male "friends" who engage in lasting relations with particular females depends on the physiology and behavior of both sexes. It is a fragile equilibrium that varies not only between species but between individuals. In primates, the "battle between the sexes" has turned into a game that allows both players to choose among various tactics, and in which the social skills are often more important than physical power.

Compared to chimpanzees, however, relations among male bonobos appear to be weak. Males tolerate one another and do not engage in the violent struggles for power that characterize chimpanzees, but alliance formation and cooperation are rare. Male bonobos seem to be ignorant of the potential power that may come from bonding with one another, but the females are not: even in the absence of close kin relationships, females have high association rates and develop complex relationships that promote mutual grooming and food-sharing and may also be used to defend food sources against other community members. Such bonding between unrelated females is considered rare both in primates and in other social mammals.

Although researchers have long been interested in the two *Pan* species' contrasting social relations, the difficulties in studying wild bonobos have hindered such comparisons. Bonobos are found only in the Democratic Republic of the Congo (DRC, formerly Zaire), the third-largest country on the African continent. They live in the lowland forests of the central part of the Congo basin, where unexplored forest extends over tens of thousands of square kilometers. The infrastructure of the country is poor, making traveling very difficult, time-consuming, and often expensive. Long-term research projects have been limited to only a few sites. Waves of political unrest moving across Central Africa have left research teams with no alternative but to leave the country. During their absence, valuable information was lost and—worse—habituated bonobos were killed by poachers. Investing in a bonobo research project requires personal commitment and financial resources that can withstand political crises and disorder. Few institutions have taken on the challenge of investing in such high-risk projects, and with good reason, even though the benefits to be expected from them make such efforts highly worthwhile.

Based on studies in zoos, bonobos have acquired the reputation of being unusually peaceful, skilfully avoiding conflicts and, if they occur, using diplomacy instead of physical power to reconcile with opponents. The social skills in the context of conflict management are paired with the elaborate and liberal sexual lives of bonobo males and females of all age classes. *Pan paniscus,* the "little" chimpanzee, has come to be stereotyped as a peaceful, sexy ape, notwithstanding that wild bonobos and bonobos in zoos sometimes have remarkably different lifestyles. No wonder that bonobos have become the darlings of the media.

Male bonobos are always vigilant, watching out for what others are doing. Adult females are the center of their universe. Equipped with a strong sex appeal, females totally absorb the interest of the males. The biological significance of strong male-female relationships appears obvious. Males who are on good terms with females are likely to enjoy advantages in terms of mating opportunities, which, in turn, will increase their reproductive success. But there is more to it than that. Males also associate with females when they are pregnant or lactating, that is, at times when no immediate benefits can be gained. Some associations may be limited to a few months, but most last for a year or longer. Temporal changes in individual relationships suggest that they are opportunistic, but we still don't know what or who terminates them and how new associations are formed. However, there is another type of male-female association, which is clearly not based on sexual interest and may last a lifetime: that of mothers and sons.

It is thought that sons benefit from this strong kinship bond because mothers have the ability to support their struggle for high status. Volker, a young male living in the Lomako forest, could serve as a textbook example of this theory. When we started our research, Volker was a shy fellow of about six or seven. At this age, female bonobos leave their natal community and join another group. In contrast, most males remain in the group into which they were born. Before becoming adult, they have to cope with a harsh social environment dominated by a few strong females and jealous males. Young males are the preferred target of aggression by older adult males, as well as by females.

Volker's position was better than those of other young males because Kamba, his mother, was the top-ranking female of the Eyengo community. When we first started observations on the group, Kamba was accompanied by two sons. Mongo, the younger, was about two years old and still in need of his mother's help in traveling and foraging. Although large enough to care for himself, Volker would not spend a single day away from his mother. Preoccupied with raising an infant, Kamba was not a particularly amenable partner for her older son, even when the little family traveled alone. And when others were around, Kamba invested in social relations with other adult females and males. While Mongo enjoyed all the benefits of physical care, Volker received indirect benefits: compared to males of similar age whose mothers had already died, he was exposed less often to aggression by adult community members, and when other adolescent males were denied entry to a feeding tree, Volker entered the patch in the shadow of his mother. Kamba also took the side of her older son during quarrels with other males. However, most of the time the strong association between mother and son appeared to be due to Volker's attachment to his mother.

During the early years of adolescence, it did not look as if his social bond with his mother interfered with Volker's other social activities. This changed when he turned nine and developed a strong interest in estrous females. Whenever Kamba traveled together with a sexually attractive female, Volker joined the other males lining up for an invitation from the female to mate. As a young male, he was often the target of aggression by older, higher-ranking males. However, in comparison to other males of his age, he showed a remarkable resilience to male aggression and did not give up in his pursuit of sex. The only thing that could deflect his interest in females was when Kamba moved away from the group. When this happened, he was obviously torn between two contrasting motivations: mating with an attractive female or following his mother. In all the cases that we observed, the latter motivation was stronger. At times when the sexual attraction seemed to exceed the bond of kinship, it never took long before he ran after Kamba, crying loudly to find her. Sometimes the lost-calls did not evoke a maternal response—leaving a sad young male alone in the forest for an hour or two until he managed to catch up with Kamba and Mongo.

Three years after we had settled into the Lomako forest, Volker had grown into adulthood: his impressively broad shoulders, covered with thick, pitch-black fur, and large testes showed that he was no longer an adolescent. While still maintaining close contact with his mother, Volker became ambitious to improve his social status among the group's males. His first target was Pink, another young male who traveled together with his mother Vanessa, an older female. With Pink, the game was easy, because Vanessa was a low-ranking outsider who lacked strong social support from other females and males. Probably because of that, Pink had fewer chances to develop social ties to other community members. When Volker and Pink displayed against each other, Kamba sometimes interfered on behalf of her son, and we had the impression that it was maternal support rather than Volker's physical strength that turned the tables in his favor. When confronted with Kamba, Pink had to back off, even when he had scored higher than Volker in a one-to-one battle. Volker's next target was Karl, an older male with a fingerless hand that limited his climbing, who tended to avoid conflicts with others. Like Pink, Karl did not show much resistance against Volker's charges. The first real challenge to Volker was Planck, an older male who occupied the second-ranking position during most of our observations. Volker's first attempts to displace Planck failed because of interventions by Max, the alpha male. It was clear that Volker would have had to wait for a couple of years to match Max's combined physical power and social competence.

Volker's big chance came when Max unexpectedly disappeared from the community. The absence of the alpha male was followed by shifting dominance relations, new friendships, and unexpected animosities. For a short time, it looked as if Planck had moved into the alpha position, and that most community members accepted the social dominance that he had developed during his long friendship with the former group leader. However, after only a few weeks, Planck's position was challenged, and Volker was the first to try his luck. First, he simply ignored the presence of Planck. He moved away when Planck presented toward him in a request for grooming. When the older male moved on a path, Volker sometimes stopped right in front of him, forcing Planck to move off the path. Then Volker started to make displays whenever he met with Planck. For several days, the forest was filled with the sounds of buttress-root drumming and the low hooting of males, signaling the arousal caused by the challenges from a new generation. When charged by the young male, Planck bowed and turned away. When Planck stretched out his arm in appeasement, Volker kept slapping and punching him. Eventually, Planck started to travel alone or at a further distance from the rest of the group, clearly avoiding the young male.

Volker had nearly reached the top position. The only male in his way was Blasius, a sturdy male with a permanently grim look on his face. Volker did not challenge Blasius, and for some time we had the impression that the two males were searching for an alternative way to solve the struggle for power by engaging in friendly interactions. However, Blasius remained vigilant, and even when Volker groomed the fur of the older male with gentle strokes, Blasius did not relax entirely. It was tempting to speculate that the two males, who were separated by fifteen years or so in age, had equivalent skills and forces, and that the social tension created by these two ambitious characters would result in shared dominance. However, instead of a happy ending, the competition for dominance status was solved by unexpected physical violence.

Volker's growing intolerance of other males was coupled with increasing efforts to maintain friendly relations with the opposite sex. Although attentive to many adult females, Volker seemed to have a particularly close and affectionate relationship with Amy, a young female who had migrated into the Eyengo community soon after we had started fieldwork at Lomako. Amy had given birth to her first infant during the same year that Volker reached his elevated position. Volker had been unusually active in courting her around the likely time of conception, and we liked to think that Amy's firstborn was sired by him. Furthermore, Amy and Kamba spent more time together than the average pair of female bonobos. Given the superior status of Kamba, the young female's affinity for the alpha female seemed to be a good way to improve her own status. Everything seemed to fit so nicely. At least that's how it would have been recorded in the natural history of bonobos if we had missed just a single day of observation.

• • •

As usual, we had left camp before dawn. To get to where the bonobos had made their nests the night before, we followed a narrow path that brought us close to the northern border of the community's home range. It was still dark when we left the trail and moved into the dense vegetation to arrive at the nest site. Most of the bonobos were still tucked into bulky bundles of leaves and twigs. The night had been cold and the bonobos had plucked extra twigs and leaves to cover themselves. Patches of sky became visible through the dark layer of canopy. Every now and then, a black body emerged from its nest. Some made new nests and disappeared into a cushion of leaves. Others hunched motionless on branches, wearing sad expressions of discomfort. After two hours of doing almost nothing, the bonobos then silently descended to the ground and moved north, only to rest for another half an hour before they decided to move again. The light was still dim and the temperature had not really changed, but the bonobos had grown hungry. Like a herd of grazing cattle, they crossed a thicket of maranthaceaes, terrestrial herbaceous plants that colonize the open areas created by fallen trees. The bonobos stopped every few steps to pull the pith out of the soft plants, each pull making a snagging sound that helped us to locate the position of the different group members in the green dark.

It was already mid-morning when we approached the Bofua River, about a kilometer from the site where the group had spent the night. The ground cover changed and the maranthaceaes gave way to small trees and vines. Suddenly, we heard bursts of food peeps and high hoots indicating that they had found a good food source. A few seconds later, the first individuals entered a large garcinia tree. The tree was not very big, but it was loaded with thousands of shiny, red, egg-shaped garcinia fruits, which are known for their unusually high concentration of sugar. We were prepared to watch a big feast and had started to take notes of those individuals entering the tree, when Volker approached, dragging a large branch behind him to emphasize his arrival. After circling around the base of the food tree, he dropped the branch, entered the food tree and moved out of our view. Next we saw Amy, who was settled in the tree, picking fruit from a branch. Her baby son clung to her belly. Little hands reached out to grab some of the red fruit. The movements were still clumsy, and the baby had a long way to go until it would be able to leave its mother. Since Amy was still young, it would take years for her to reach a high dominance status. Accordingly, it would not be easy for her son to reach a high-ranking position within the community. However, unlike the sons of females older than Amy, who are not likely to have maternal kin around when they reach adulthood, her son had good chances of enjoying the benefits of maternal support in the distant future.

Amy had finished her snack but remained at the same spot, watching her son with affection. The next things happened so quickly that we could only reconstruct the sequence of events by merging the observations of two people, some recorded on an audiotape. Dense vegetation, the flurry of black bodies, and the speed of the attack prevented detailed accounts of the behavioral interactions.

> Volker jumps on the branch that holds Amy and her baby. For a second the female seems to lose balance but then maintains a firm grip and pushes Volker off the branch. The male jumps to the ground and is followed by a screaming Amy. The descent of Volker and Amy initiates a rush, as other adult females and males drop out of the tree and within seconds the forest transforms into a battle ground. Details are camouflaged by the dense vegetation, but the frightening noise of screaming bonobos indicates that this is not a mock fight but a fierce struggle. We hear breaking branches and the hollow sounds of what seem to be hard slaps on a body. For a second, we see three individuals dragging Volker by one leg. The male looks helpless, trying to hold on to whatever he can grab. Then the cluster of bodies disappears into the dark shelter underneath a fallen tree. We follow, but alarm calls directed toward us from other bonobos make us stop our

> attempts to get a closer look. It is difficult to control the natural urge to rescue a victim, but the last thing we want is to interfere in the domestic affairs of the bonobos. We see black bodies appear for seconds before they are swallowed up by dense vegetation. For a moment, Blasius moves out of the bushes. His erect hair makes him look much bigger than he is. When he sees us, he stops, breathing heavily. Then he barks at us, breaks a tree, and drags it across the mutilated vegetation. The faces of the bonobos that we have known so well for several years now reflect emotions that we have never seen before. Others move in and out of view, but not the one we hope most to see. Then, for a few seconds the curtain lifts when a male—it must be Pink—bends a small tree away. Volker is sitting on the ground, holding on to a tree trunk with hands and feet, the screaming face turned toward an enemy that must be so familiar to him.

This is the last picture that we have of Volker. Half an hour later, when the forest reverted to a scary silence, when males and females returned to the garcinia tree, Volker was gone. Like a spirit, he had disappeared, leaving us to speculate about his fate.

Our field notes revealed that fifteen or more adult individuals had participated in the battle. Many other details of this event remain unknown to us, but one important fact was clear: the only adult individual who stayed away from the attack was Kamba. Knowing that Volker and Kamba were still inseparable, we frantically searched for the mother of the victim. When we finally found her, she was hiding high in the canopy of a tall tree, almost invisible—alone. How could this have happened? Kamba, the dominant member of the Eyengo community had not tried to intervene when one of her children was in danger? Why had she not stopped Amy, a young, low-ranking female and close friend, from charging her high-ranking adult son? What was it that turned the young mother into a fierce fighter? Why did the community respond in unison, attacking one of its own members? There seemed to be no logic to this. It must have been an accident, an unusual case where everything went wrong. A perfect storm.

When we set out to study wild bonobos, we were prepared to see unusual things. But what we had observed was not only unusual—it was disturbingly unexpected. Our preconceived notions of bonobo behavior were suddenly shattered. The attack on Volker by the other bonobos in the group, who had known him since his infancy, shattered the cherished image that had emerged from so many studies. It was like looking through out-of-focus binoculars. There is a strong urge to adjust the world around us to an image based on experience, based on expectation. Instead of satisfying our sense of academic curiosity, this event made us feel unhappy, because we did not like the cruelty that we had seen and had no good explanation for it. Not yet.

When the bonobos finally left the garcinia tree, we followed them until they made their night nests. All the individuals who had been together in the morning were still present. Only Volker was missing. The next day, we returned to the battlefield in search of traces: a dead body, blood, or other signs that could be used to reconstruct the events that had been hidden behind the leaves. All we found were some small bundles of black hair. For a year, we looked out for Volker, but our data sheets remained empty; there is no entry for Volker.

Male bonobo males do sometimes transfer into other social groups, but we also know that the acceptance of a strange male by members of other communities is uncommon. When groups encounter a strange male, the response is hostile and involves intense physical aggression that is rarely seen in bonobo communities, with both males and females violently attacking the stranger. In the absence of direct evidence, we can only speculate about Volker's fate: in the best case, he survived the gang attack and was accepted by another community. In the worst case, he moved away from the spot where he was last seen and died as a result of his injuries. This story, some pictures, and a genetic fingerprint are what remains.

In cases like this, one's feelings are unreliable, and it was days before we were able to analyze the event with a sober mind. After discussing various scenarios, we were left with intriguing puzzles. Why did Amy risk attacking Volker? And why did Kamba not come to the defense of her son? Amy's behavior indicates that she considered Volker's behavior to be unusually dangerous. Did he intend to commit infanticide? Male infanticide—eliminating the offspring of another male—has been reported from many different primate species. The killing of an unweaned infant leads to the mother's resumption of estrus, and the male thus increases his chances to reproduce. Although infanticide has never been reported in bonobos, its absence has not been taken as a sign of the friendliness of males to infants, but has been ascribed to the strength of bonds between females. Following this line of thinking, one should not be surprised to see, now and then, a male that is ready to commit infanticide. Was this what we had seen? We'll never know for sure. But circumstantial evidence came later when genetic analyses revealed that Amy's son had been sired by an unknown male who did not belong to the Eyengo community. While this information came as an absolute surprise, it offered an explanation for something that otherwise looked like a bizarre case of intracommunal violence.

Amy had an impressive record of mating with males from the Eyengo community, but she must have slipped away from the group just when the likelihood of her fertilization was very high. From this perspective, the male who was most ambitious to become the new leader of the Eyengo community had a good reason to kill Amy's son. Even if we accept this, however, we are left with the second puzzling question of why Kamba did not attempt to rescue her son. How did it happen that a mother did not defend her offspring against a squad of charging males and females? Was this the price of female bonding? Did the wisdom of nature force bonobo females to punish a male who tried to commit infanticide, even if he was kin? Could water be thicker than blood? This unexpected case of communal aggression against a resident male bonobo, which calls into question both the strength of kinship bonds and the image of *Pan paniscus* as the peaceful ape, shows that we are still far from understanding the social world of bonobos and the extent to which they may confound our expectations.

FURTHER READING

Furuchi, T., and J. Thompson, eds. 2008. *Bonobos: Behavior, ecology, and conservation.* New York: Springer.

Hohmann, G. 2001. Association and social interactions between strangers and residents in bonobos (*Pan paniscus*). *Primates* 42:91–99.

Hohmann, G., and B. Fruth. 2000. Use and function of genital contacts among female bonobos. *Animal Behaviour* 60:107–20.

———. 2003. Intra- and intersexual aggression by bonobos in the context of mating. *Behaviour* 140:1389–1413.

Surbeck, M., and G. Hohmann. 2008. Primate hunting by bonobos at Lui Kotale, Salonga National Park. *Current Biology* 18:R906–7.

Waal, F.B.M. de, and F. Lanting. 1998. *Bonobo: The forgotten ape.* Berkeley: University of California Press.

FIVE

Our Cousins in the Forest—or Bushmeat?

CHRISTOPHE BOESCH

It was May 1987 in Côte d'Ivoire's Taï forest, and warm sunshine penetrated the dense vegetation. I was enjoying a quiet walk with seven chimpanzees toward the northern part of their territory. Darwin, Brutus, Ulysse, and the others had adopted a leisurely pace so as to inspect all the *Nauclea diderrichii* trees we passed. The huge nauclea trees with their high, straight trunks and amazingly hard wood, which even termites can't eat, are very popular with logging companies, whose roads cut through the pristine forests for miles. But here within the boundaries of the national park, chimpanzees could enjoy the rich, fleshy, juicy nauclea fruits covering the ground during the fruiting season, filling their bellies as they moved from one tree to another. Chimpanzees are very fussy eaters when it comes to fruit and eat it only when ripe. Even then, they prefer to peel fruits such as nauclea and chew the flesh to extract the juice, after which they spit out the remaining pulp and seed in what by primatologists call a "wadge." Each time they came upon a new fruiting tree, they gave excited food calls and quickly fell to eating. The small forest antelopes that also like nauclea fruits would swiftly run away to avoid these noisy newcomers.

At 9:50 A.M., Darwin and friends were enjoying another good nauclea feeding session. I was sitting nearby, looking at the abundant birdlife and listing to the monkeys calling in the forest around us, when suddenly I was deafened by a terrible BANG that made my heart turn over. What was that?? It took me a long quarter of a second to realize that someone was shooting at us from very nearby. The chimpanzees were being shot at! We were being shot at! I was being shot at! Someone was trying to kill them and me! This was a very strange new sensation. I heard a chimpanzee noisily running away on the ground, as if wounded and clutching at saplings to avoid falling down. Who was it? Would he die? What could I do when there was a man with a gun hidden near me? I saw Ulysse and Brutus stand up for half a second and look toward the direction where the gunshot had come from and then silently run away. As a reflex reaction, I followed them for thirty-five meters, but then I stopped, saying to myself, "What if the poacher attacks the chimpanzees again? I have to prevent that, and he needs to know a human is here!" Hidden by the forest, I silently walked back a few meters and then shouted twice, "You are in the middle of the national park; it is illegal to shoot animals!" Words against a gun! I hoped the sheer sound of a human voice would scare him away.

After twenty minutes of waiting in silence, I tracked back to where we had been shot at and saw that the poacher had been only twenty meters away from us, because there were signs on the ground indicating where he had knelt down and shot at the chimpanzees. He could have killed me as well. An empty shotgun shell of Liberian origin was lying nearby. I did not find any further traces of the poacher or of a killed or injured chimpanzee. Two hours later, when I caught up with Brutus and Ulysse two kilometers further to the south, all the others were there except for one. Where was Darwin? For the next three months, he was missing. Had he been killed by the poacher? Or was he badly injured and unable to rejoin the group? With such questions in your head, three months is a long, very long, time. After two months, I gave up hope of ever seeing Darwin again.

Unexpectedly, however, Darwin reappeared in the group three months later. What a lovely surprise! When I first saw him, I tried to see if he had sustained any injuries during the poacher's attack. Faithful to his reputation for being a survivor, Darwin appeared to have survived the attack without any damage. Years before, when I had first identified him, I was impressed to see this healthy young male chimpanzee vigorously climb a large tree, joining his voice to the chorus of hungry, excited chimpanzees feeding above him. However, there was something very strikingly odd about his way of climbing: he was placing his left knee against the tree trunk and not his foot, making it an unbelievable challenge to climb a vertical trunk with so much weight to support with his arms. His left foot was totally missing, probably due to a bad deformity at birth, while his right foot seemed broken in the middle, so that he could not hold on to anything with it. These handicaps limiting his ability to climb trees had probably caused some periods of dire malnutrition when he was a youngster, and as a consequence, he had very bad teeth. But what a living testimony to the fight for survival!

I wondered why Darwin had gone missing for such a long time after the attack, something most unusual for a strongly social adult male chimpanzee, especially since he did not seem injured at all. In Darwin's case, we'll never know for sure, but the most common reason for a male to be absent from the main group for a long time is sex. Chimpanzees have a "fission-fusion" grouping pattern, which means that not all group members are always together. At any given time, there are many subgroups, or "parties," and adults may also spend time alone. Male chimpanzees regularly leave the main group to consort with a sexually active female for many days or weeks. In this way, they increase their certainty of being the father of the resulting infant. With his physical handicaps, Darwin had limited possibilities of climbing the social ladder of the male dominance hierarchy, and, during the twelve years I saw him as an adult, he remained one of the lowest-ranking males. However, that did not prevent him from having a strong interest in the females, and genetic studies revealed that he was successful in siring two sons. His second son, Nino, became the dominant male of the community after the death of his father and he sired many infants, assuring the future representation of Darwin's ability to fight for survival. This illustrates how even handicapped individuals can be successful in perpetuating their genes in future generations.

MURDER IN THE FOREST

Darwin's disappearance brought home to me the importance of explaining to the villagers in our region what we were doing in the forest. I asked the numerous young men from those villages who worked for

THE HUNTING OF GREAT APES FOR MEAT

Traditionally, most human populations living in forest regions of Africa have recognized that chimpanzees are closely related to us, and some therefore treat them as "totem"—it is forbidden to kill or eat them. This is often justified by a story where in the past the chimpanzees contributed positively to the survival of a family member or to protecting the village. In other cases, people believe that chimpanzees have some special force or talent that can be acquired by eating their bones, hands, or parts of their heads in special traditional ways. They may attribute some magical therapeutic properties to chimpanzees; for example that young babies will grow strong and healthy if they have a bath in chimpanzee bone powder. As a result, chimpanzee meat is considered a special treat to local human populations in many regions in Africa, and that makes them especially vulnerable to hunting. While in general chimpanzees are not the target of poachers, because they are so difficult to find and approach, if hunters do spot them they will try their luck and kill them if possible.

Hunting for bushmeat has increased with the introduction of automatic weapons in Africa, with the increase of human populations, and with the opening up of the most inaccessible forest regions of Africa by logging companies. As a result, no wild animals or great ape populations are safe from hunting. This hunting has reached tragic ▸

us to explain to their friends and families that poachers were putting them directly at risk when they attacked our study animals, because they might accidentally shoot a person. However, as everywhere in the world, there are those in Côte d'Ivoire who don't respect nature or don't want to follow the rules of national parks. This was making our work difficult, and I kept indicating my presence in the forest whenever I was aware of the presence of poachers by shouting and drumming on tree buttress roots. For a long time, I felt that things had quietened down,

▸ proportions and is now called the "bushmeat crisis," because the products of the hunting are sold in big cities and this has become a very profitable economic sector. This has led to the "empty forest syndrome" in which stretches of pristine forests are found, but are devoid of any animals, because the demand for bushmeat has wiped them out. All animal species suffer, but with their especially slow reproductive rates, both chimpanzees and gorillas are particularly at risk. Since chimpanzee females typically produce an infant only once every five years, chimpanzees are especially slow to recover from repeated decimation by hunting or destruction of their habitats. While some studies have shown that only about 3 to 4 percent of bushmeat is from great apes, that is enough to produce a general decrease in their populations in most of their range throughout Africa.

Effective prevention of the hunting of great apes and other endangered animals has proven very difficult, particularly in forests where hunters can operate elusively on foot. Recent work in the Taï National Park in Côte d'Ivoire and in the Virunga Volcanoes region of Rwanda, the Democratic Republic of the Congo, and Uganda has shown that in addition to traditional law enforcement, properly managed ecotourism and long-term, continuous, and supportive scientific research projects are the best ways to protect animals from illegal hunting in protected areas.

and that once our presence was accepted, the poaching problem would disappear. How naïve I was!

In the pitch-dark of early morning on September 1, 2004, Ferdinand, carrying a gun, and his little assistant, Lucien, entered what they knew to be the research area of the Taï chimpanzee project. People in the region were becoming aware that hunting was much more profitable in this area where researchers were working, because they had depleted much of the monkey population in other less protected regions of the park. A poacher could kill six to seven monkeys in one day in the research area, while in other regions of the forest, he might kill only one. There clearly was a risk of being recognized by one of the many assistants working for my project if caught poaching in the research area, but Ferdinand and Lucien were ready to take this risk. Léon, the owner of the gun, who had given them the "contract" to hunt, had ordered them to bring back some monkeys for a party that night.

This morning, they were lucky. They heard chimpanzee calls nearby before the dawn broke. Previously in the village they had heard that the chimpanzees in this region were not afraid of humans, and they hoped to be able to kill one of them. True, the villagers also said that chimpanzees are almost like humans, using tools, making war, acting as a cooperative team, and adopting orphans. All these amazing chimpanzee behaviors had been explained when a local conservation group gave a theatrical performance in the local village. This really helped explain the mystery of the chimpanzees that live in the forest. As in many places in Africa, even though people may live very close to the forest, they don't have much of an understanding of the wildlife and tend to view it simply with fear or as an opportunity for hunting. But this morning, Ferdinand and Lucien had to bring some meat back to Léon—their job was more important than stories.

As they slowly drew closer to the chimpanzees' vocalizations, they were lucky again. They saw a big chimpanzee quietly looking at them from a low branch. Without hesitation, Ferdinand aimed at him and shot Bésar, a 15-year-old orphan male, in the head. Lucien quickly put the dead chimpanzee on his back, and they started to walk back to the village. They had accomplished their goal before daybreak.

Then their luck abandoned them: after walking only a hundred meters, Ferdinand was bitten by a snake. He was able to kill it with his bush knife, but he immediately felt very weak and dizzy. He fainted before he could alert Lucien to what had happened. Lucien had been walking in front of him, and it was only after a while that he realized that Ferdinand wasn't with him any more. He called for him twice and didn't get a response. Keen to leave this part of the forest before the researchers arrived, he rushed back to the village with the dead chimpanzee on his back. Once the story spread around the village, Ferdinand's father was furious that Lucien had abandoned his son in the forest, and he threatened to kill all of Léon's children (since Léon was the one who had sent them into the forest to hunt) if Ferdinand did not come back alive. Knowing that Ferdinand had been in the park illegally and, more specifically, in the research area, it was difficult for them to mount a full-blown rescue operation. As a result, the search parties made slow progress. Two days later, a very frail Ferdinand came back to the village. A party was organized to celebrate his survival, during which the chimpanzee Bésar was eaten.

At first we were all shocked by the unexplained disappearance of Bésar. What could have happened to him? One of my students, Emmanuelle, had already started telling people in the nearby villages about the terrible consequences poaching could have for her research project and, more important, about the huge threat it posed to the long-term survival of the chimpanzees. We were happy to hear that many people wanted to support the chimpanzees and our research project, because it was providing the only jobs in the region. Because Emmanuelle had this ongoing dialogue with the villagers, it was through patient discussions that she learned of the expedition during which Ferdinand and Lucien had killed Bésar.

BEING AN ORPHAN IN THE FOREST

When I first saw Bésar in 1999, he was six years old, and his mother was still alive. Sadly, she died when he was only seven, and he had no older siblings to adopt him. For protection from the many squabbles in this large chimpanzee community, he constantly followed the big dominant male of the community, Zyon. As with many dominant males in the Taï forest, Zyon was very tolerant and sometimes even protective of male orphans. He was regularly seen with his clique of young followers, Bésar and two other younger male orphans. These young males would accompany Zyon wherever he went, and they thus learned some of chimpanzees' most masculine activities at an early age, including some of the risky ones.

I vividly remember one day when Bésar was eight years old. The group was surprised by calls from the neighboring community, in the southern boundary area shared by the two groups. All five adult males rushed ahead to face the intruders, and I ran past the females, who were trying to keep up with the males. The females were very noisy as they followed the males, drumming frequently on the big buttress roots of large trees to give the false impression that the number of male chimpanzees approaching was much larger than it was, because normally males do most of the drumming. Then I caught up with lonely Bésar. He was screaming from fear, but at the same time trying to catch up with Zyon, who was leading the counterattack. I followed Bésar, behind the males, who were trying to locate and chase the intruders, and ahead of the females, who continued to bluff the intruders about the real strength of the attack with their drumming. While I knew that those coming behind were only females, the sound of them progressing in such a noisy and coherent fashion was quite impressive. Together Bésar and I ran a few hundred meters and caught up with Zyon and his four male buddies, all of us listening silently and carefully to the progression of the females, while trying to locate the intruders. Bésar immediately went to groom Zyon, in an attempt to calm himself down.

We were by now within the limits of the neighboring community's territory. Suddenly from the south, we heard additional drumming from the intruders. Zyon and another male quickly embraced one another and then without any hesitation the five adult males ran silently toward the opponents. Whimpering softly, Bésar stayed behind, possibly aware that it was by now much too dangerous for a little guy like him to be part of such a commando assault. I tried to keep up with the adult males, but a tall bipedal human is much clumsier in a dense rain forest than a strong ball of fur running quadrupedally. Accepting my handicap, I tried to locate them from their calls. But the fight turned out to be only a wild chase, with males running in many directions without any direct physical contact, and the action was short-lived.

Left alone, I went back to Bésar, who had joined the females. Now they were quieter, eating some fruit while listening to the chase ahead. From time to time, they would give support barks when the calls got louder from the males. After ten minutes, Zyon, as the oldest male, came back first and greeted the females by displaying past all of them. Bésar immediately went to groom him. It took twenty more minutes before the younger, more enterprising males joined the rest of the group, and then all of them headed back within their territorial boundaries.

Like all orphans, Bésar definitely grew more slowly than other youngsters, but after a slow start he caught up nicely. At the time of his death, he was comparable in size to any of the community's other adult males. He grew into a skillful user of tools for nut cracking and a talented cooperative hunter. Then the poachers came to kill him. What an absurd, criminal waste of a life for just one meal!

Emmanuelle found many good friends of the chimpanzees among the local population, however, and with their help and that of the local authorities, Léon was brought before the local court and received an eighteen-month jail sentence for having organized the killing of chimpanzees in the Taï National Park. In the meantime, Ferdinand had fled to Liberia to avoid facing justice.

Darwin and Bésar represent two extremes of the bushmeat problem so widespread in Africa. Darwin's case shows that the chimpanzees can fare well if we give them a chance. On the other hand, Bésar's story shows that it takes only one incident to remove a member from a community forever.

Killing wild animals for meat has always been a common practice over large parts of tropical Africa for many different reasons, especially as a source of animal protein and as "medicine" in traditional pharmacies. Statistics and gruesome photos alone don't illustrate who those animals were and what their lives in their community and the forest had been. Darwin and Bésar are only two of the thousands who have had to face guns. Female chimpanzees reach maturity at fourteen years of age, and then they produce an infant only every five years. Thus it requires twenty years to replace a Bésar, time during which poaching is taking its toll on even more chimpanzees on a daily basis. This explains why great apes, with their slow reproductive rates, are so sensitive to poaching and why some large forests can simply be emptied of apes in short order.

What does the future look like for the chimpanzees? Bésar's death has taught me to be careful about making predictions. Zyon's young clique of males eventually dissolved as they grew up. Five years later, Zyon was an old male of the group and was seen more often by himself. In the meantime, however, three young males had reached adulthood and added their strength to that of Zyon and the other males, forming a new force in the forest, which we saw successfully challenging neighboring groups. For the first time in years, three young females were seen to immigrate into the group. In chimpanzees, females transfer between groups when they reach maturity, and they seem to select groups with numerous males. Thanks to our presence, poaching can perhaps be controlled, and by allowing males to live, the long-term survival of groups of chimpanzees might be possible.

FURTHER READING

Boesch, C., H. Boesch, Z. Bertin Goné Bi, E. Normand, and I. Herbinger. 2008. The contribution of long-term research by the Taï Chimpanzee Project to conservation. In *Science and conservation in African forests: The benefits of long-term research,* ed. Richard Wrangham and Elizabeth Ross, 184–200. Cambridge: Cambridge University Press.

Boesch, C., C. Gnakouri, L. Marques, G. Nohon, I. Herbinger, F. Lauginie, H. Boesch, S. Kouamé, M. Traoré, and F. Akindes. 2005. Chimpanzee conservation and theatre: A case study on an awareness project around the Taï National Park, Côte d'Ivoire. In *Conservation in the 21st century: Gorillas as a case study,* ed. T. Stoinski, P. Mehlman, and D. Steklis. New York: Kluwer Academic and Plenum Publishers.

Campbell, G., H. Kuehl, P. N'Goran, and C. Boesch. 2008. Alarming decline of West African chimpanzees in Côte d'Ivoire. *Current Biology* 18:R903–4.

Köndgen, S., H. Kühl, P.K. N'Goran, P.D. Walsh, S. Schenk, N. Ernst, R. Biek, P. Formenty, K. Mätz-Rensing, B. Schweiger, S. Junglen, H. Ellerbrok, A. Nitsche, T. Briese, W.I. Lipkin, G. Pauli, C. Boesch, and F.H. Leendertz. 2008. Pandemic human viruses cause decline of endangered great apes. *Current Biology* 8 (4):260–64.

Morgan, D., and C. Sanz. 2007. Best practice guidelines for reducing the impact of commercial logging on wild apes in West Equatorial Africa. Gland, Switzerland: IUCN/SSC Primate Specialist Group (PSG).

SIX

Discovering Chimpanzee Traditions

CRICKETTE SANZ AND DAVID MORGAN

There was both concern and excitement in our hushed voices as we confirmed that neither of us recognized this young female chimpanzee who was calmly peering down at us from the low branches of a towering fig tree in the Congo River Basin on a sultry afternoon in May 2002. Dave flipped through his field book, but none of his sketches depicted the graceful features of this female's light face. She glanced toward the other chimpanzees in the canopy as if to gauge their response to the strange party of bipedal apes who had just arrived with their backpacks, binoculars, and cameras. Macallan, a burly male and dominant figure in this group, was preoccupied with fitting as many figs in his mouth as possible before starting to chew them into a manageable mouthful. An older female named Moja was smacking her lips in concentration as she used both hands to part the hair on her youngster's head. She seemed oblivious to us while she dexterously used her lips to remove forest debris that he had picked up tumbling through the undergrowth with his playmates. Another youngster climbed toward us to gather some figs directly overhead. Seemingly reassured by the others' calm behavior, the unknown young female turned away and continued to gather plump yellow figs from the large canopy of the ripening tree. Although the chimpanzees did not show any noticeable response to us or the young immigrant, we knew that this first meeting was a landmark event for our newly established site for studying these elusive apes.

We had been anticipating the occasional dispersal of young females from the Moto community and arrivals from other groups, because it is most often the females who transfer between chimpanzee groups. We had documented the mysterious disappearance of some younger females who had just become sexually mature and were likely candidates for transfer, but we could not confirm their fates. Now a new young female had been found among the familiar faces of our main study group. After encountering her several times during the next week, we decided to name her Dorothy after the lost girl from the Wizard of Oz. Her status as a new arrival was further confirmed by the behavior of the other chimpanzees, and it was clear to us that she was trying as diligently as we were to gain the acceptance of the Moto community.

It was a rainy day in February 1999 when Dave first trekked into the remote forests of the Goualougo Triangle. He had already spent two years in Nouabalé-Ndoki National Park in the northern Republic of the Congo studying gorillas at the Mbeli Bai clearing. From the vantage point of a tree-house overlooking the swampy clearing, he had helped identify more than a hundred gorillas that waded into the inundated forest opening to forage on herbs and grasses (see chapter 8). After proving his staying power in these remote forests, Dave's lifelong dream came true when he was asked to estimate the number of chimpanzees in an area called the Goualougo Triangle, which is located five hours' walk down a footpath from Mbeli in the midst of Congo's lowland forest.

The goal of Dave's initial mission was to use his experience of identifying apes at Mbeli to census the chimpanzees and document their responses to humans, information that would be used to protect these apes and their forest home from future timber exploitation. Chimpanzees in this particular forest had a local reputation for being "naïve" to humans. Village elders who had trekked extensively throughout the northern Republic of the Congo reported that the chimpanzees in this area would approach them curiously and even gather around the humans to stare at them for hours. They also reported seeing them using large clubs to pound open beehives to gather honey, and sticks to puncture termite nests so as to fish out the insects. All wild chimpanzees use tools, but the composition of their particular tool kits differs between populations, because they invent and use tools relevant to the challenges in their particular environments. Since very little was known about great apes in this region, the Goualougo Triangle seemed like an ideal place to study wild chimpanzees. Dave knew that this was a unique opportunity to gather insights about chimpanzees of the Congo Basin while also contributing to their conservation. There were several long-term studies of chimpanzees in East and West Africa, but a curious paucity of research in the dense lowland forests of central Africa.

Several established scientists had made valiant efforts to study chimpanzees in the Republic of the Congo, Gabon, and the Central African Republic, but none had succeeded in identifying an entire group and consistently following them through the forest. There was speculation that the high density of elephants and gorillas had made habituation of chimpanzees a more complicated undertaking in this region. It was also coming to light that the bushmeat trade was a widespread, commercially profitable business in central Africa. As unthinkable as hunting chimpanzees and gorillas for meat was to us, it was easily confirmed by the fearful responses of apes to humans and in casual conversations with local villagers who had hunted them. Wild apes that have been hunted tend wisely to remain silent or to immediately flee when detected by humans. Fortunately, this was not the case in the remote forests of the Goualougo Triangle, which had been naturally protected by the two large rivers that form its eastern and western boundaries. The large swaths of swamp on either side of these rivers had discouraged even the most determined hunters. In contrast to the apes on the wrong side of the river, chimpanzees in the Goualougo Triangle often confidently asserted their ownership of these forests. Instead of retreating from humans, they generally showed an interest in making their acquaintance.

TOOL USE AND TRADITIONS

Humans constantly take advantage of complex tools. Saws and hammers have helped build the homes that protect us from the elements. Automobiles transport us from place to place. Personal computers and the Internet have enhanced our abilities to store and communicate information. A tool can be simply defined as anything used to attain a goal or change one's environment. Using objects to protect yourself or obtain food is biologically relevant in that it can increase the probability of survival and production of offspring.

Given that we humans are very good at using tools to get what we want, it is easy to see why tool use was until recently considered a unique characteristic of "man the toolmaker." This perception changed, however, after the first observation of tool use by wild chimpanzees was reported by Jane Goodall in the early 1960s. In the Gombe forest of Tanzania, Goodall spied an adult male ape threading a slender grass stem into an earthen termite nest and then carefully extracting the stem with insects clinging to the tool. Since her pioneering observations, field researchers at other sites have subsequently shown that all wild chimpanzee populations exhibit some form of tool use.

Regular tool use has now been documented in various animal species, ranging from crows to capuchin monkeys, but complex tool use in natural settings seems limited to chimpanzees and orangutans. Chimpanzees in particular have a relatively large tool kit that contains a variety of different objects. Also, these apes have shown advanced skills in using multiple tools to achieve their goals.

Tool-using skills passed from one generation to the next through social learning constitute technological traditions. Most of the tools chimpanzees use are made of perishable materials that eventually decay and are thus undetected in the archaeological record. However, some populations of chimpanzees also use stone tools, and such use has been traced back more than 4,000 years in the archaeological record. Although tool assemblages can provide valuable insights into ▸

▸ the abilities of their makers, it is preferable to observe the tool users in action so as to study how they make tools, use them to accomplish their goals, and transmit this information between individuals in their social group. Extant tool-using ape populations provide an opportunity to study such technological traditions and their importance within the social culture of each ape population, which is also relevant to understanding the evolution of tool use in our own species.

Ecological conditions are likely to have led to the emergence of tool use. Each habitat provides unique foraging challenges to its residents. Tools can be used effectively to access otherwise inaccessible resources, such as honey in a beehive or seeds that are well protected in a hard neesia pod lined with stinging hairs (done by orangutans). Some birds use tools on a regular basis, but most of these behaviors are preprogrammed adaptations that emerge even when the birds are raised in isolation from others. In contrast, the tool use of chimpanzees and orangutans seems to be more insightful and is acquired through information gathered during the prolonged juvenile period, when they are in almost constant association with their mothers. By the time they are mature, these apes can use objects in creative ways and apply tools to different situations than those for which the tool was originally designed, both of which are hallmarks of intelligent tool use.

For their tool-using skills to be preserved over time, this information must be passed from one individual to another through social exchanges. Laboratory studies have shown that chimpanzees can transmit tool-using skills within and between groups. Further, individuals within these groups tended to show conformity in their tool using with other individuals in their group, even when they were aware of other techniques to accomplish the same goal. Although it is more difficult to study the processes of social learning in wild ape populations, there is clear evidence that chimpanzee mothers play an important role in the acquisition of tool-using behaviors by their offspring. Research has shown that mothers will change their behavior to provide tool-using opportunities for their offspring, even if this lowers the mother's ▸

It had taken more than six months of tireless tracking to meet all of the members of the Moto community. It is simply not possible to make a head count of an entire group during a single encounter, because at any given time the community comprises several "parties," which range throughout the group's territory; all are scarcely ever in the same place at the same time. These parties may merge and split several times a day in response to the abundance of food, number of sexually recep-

▸ foraging efficiency. For example, a chimpanzee mother in the Bossou forest in Guinea, West Africa, will choose to dip for army ants traveling on the ground in the forest rather than at the ant nests, which are more dangerous for young tool users. Her own dipping yield would be higher at the ant nests, but she assumes the cost of lower foraging returns to provide a safe learning opportunity for her child. In the Taï forest in Côte d'Ivoire, mothers share their hammers with youngsters learning to crack open nuts, a skill that characterizes their chimpanzee society.

Most conservation efforts monitor success by counting the number of remaining apes in an area or population. In addition to studying population size and distribution, it is important to determine whether the traditional skills of wild apes are also in danger of extinction. Most of the world's remaining great apes are living in environments that are rapidly changing with the advance of mechanized logging and human settlement. The tool-using traditions of chimpanzees in equatorial Africa and orangutans in Indonesia are not likely to withstand the myriad of negative impacts that are often associated with the arrival of their modern human relations. For example, the termite nests where chimpanzees learn and practice their tool-using skills are often destroyed during timber extraction in the northern Congo. Much of this collateral damage to the forest is avoidable and could be prevented by stricter forest regulations and enforcement of protection measures in ape habitats. It is thus our responsibility to use all of the tools at our disposal to preserve remaining ape traditions.

tive females, or a myriad of other factors. It was not unusual for the forest suddenly to erupt with hoots, screams, pants, and food grunts as a party of chimpanzees celebrated their arrival at a large fruiting tree. Such a food bonanza could attract members from distant reaches of the Moto community's range and were a great reward for our hours spent searching the forests for these apes. On other days it could be much more difficult to locate a small party of chimpanzees who were quietly feeding on smaller and less desirable foods. To determine the membership of a community, we needed to find a cross-section of such parties and document all the social links between individuals.

The lack of human pressures in the Goualougo Triangle had certainly given us an advantage in winning the trust of chimpanzees in the dense lowland forests of the Congo Basin, where chimpanzees have been notoriously difficult to identify and observe. In contrast, Dorothy would have to endure much more intense trials of initiation to become a member of the Moto community, some involving intimidation and aggression from other chimpanzees. The dominant males took no interest in her, even when her sexual receptivity was immodestly advertised by the large, bright pink swelling of her genitals each month. Her approaches to several of the resident females were often rebuffed or politely tolerated at best. She developed the alternative tactic of befriending their juvenile children. These kids seemed quite willing to accept anyone who might be interested in play, but Dorothy had to be careful, because the slightest whimper or squeal of a discontented youngster would bring a mother's wrath.

One sunny afternoon, a small party of mothers and their children were relaxing after having gorged themselves on strychnos fruits, which have a sweet fragrance reminiscent of potpourri. Dorothy was doing her best to groom a juvenile male named Leakey. Grooming is a common tactic among chimpanzees to build social bonds, but Leakey's reciprocation involved only pulling on Dorothy's hair and playfully grabbing at her fingers. She lowered her head and concentrated on an unruly tuft of hair on Leakey's leg. He slapped her head with both hands and chuckled.

Dorothy continued to groom, disappointing the young male, who was used to getting his way. We could see that a tantrum was brewing and as soon as the first cry left Leakey's lips, Theresa was on her feet and bounding toward the pair to rescue her son. The experienced mother whisked Leakey behind her and then set her sights on Dorothy. The young female screamed and cowered away as the older female started hitting and biting her. Dorothy recklessly backed down the branch, which bowed under the weight of the two chimpanzees. Theresa relentlessly pursued Dorothy until the younger female scrambled without pause into an adjacent tree. Dorothy was still screaming and crouching, as Theresa returned to her resting place with Leakey in tow. The young female's high-pitched shrieks continued for several minutes, but it was as if she were invisible to the other chimpanzees in the group, who did not offer her as much as a sideways glance in solace.

We often wished that we could be invisible in the presence of these chimpanzees so as to more closely watch the intimate details of their lives. It takes several years to fully habituate a group of wild chimpanzees to the presence of human observers. There are no field guides or instruction manuals on how to begin a study of wild apes, but we had read all about Jane Goodall's arrival at Gombe, Toshisada Nishida's research at Mahale, and Christophe Boesch's studies of the chimpanzees in the Taï forest. These scientists were the guardians of the ape legacy, and it was clear from accounts of their early years that the tricks of the trade would come only with experience. We desperately tried to imitate our mentors in recording the behavior of the chimpanzees with composure and scientific precision, but this seemed nearly impossible when trapped in a tangle of thorny vines, chased by a swarm of wasps, or charged by an angry elephant. We could only hope that it was normal for the first months of research to seem more fitting for an adventure magazine than a scientific journal.

If we could only win their trust, it was likely that the Congo Basin chimpanzees could show us unique social customs and tool-using practices that had never been documented elsewhere, because each community of chimpanzees has a unique culture. It is thought that such behavioral traditions are preserved over several generations through social transmission between individuals. We often found bits of vegetation near large termite nests with one end fashioned into what looked like a green paintbrush. The chimpanzees had used these tools to "fish" termites from their nest by inserting the brush into the nest and withdrawing insects that attacked the invading tool. The chimpanzees had wisely chosen to use a tool instead of their hands to gather the termites, whose soldiers sport large, razor-sharp mandibles. Rather than use a tool, one could break the termite nest open with brute force, which is the technique favored by gorillas, but this often results in destruction of the nest. In contrast, the chimpanzees' technique of using thin stalks to extract termites does not destroy the termite mound, which harbors a protein-rich food source and can be revisited for years to come.

We collected and studied the tools that chimpanzees had discarded after use at termite nests or beehives, but to make any solid contribution to scientific literature we needed observations of the apes actually using the tools. Elephant and gorilla researchers working in the northern Republic of the Congo reported glimpses of chimpanzees using large clubs to break into beehives to gather honey, which had not been documented in eastern or western chimpanzee populations. There were also intriguing reports of more than one type of tool being used by chimpanzees in gathering honey from beehives and extracting termites from their nests, referred to as a "tool set." A tool set might consist of a stout wooden stick to open an insect nest and then a slender stalk to extract the insects without getting bitten.

The use of tool sets raises intriguing questions about chimpanzee cognition. Do they have distinct mental templates for each of these types of tools, and, if so, do they prepare the second type of tool in advance of arrival at an insect nest? Tool sets have rarely been observed in wild apes and could represent a more complex technology than that previously attributed to them. The local lore about chimpanzee tool use and traces of tools that we recovered at insect nests were all that was needed to convince us that the chimpanzees of the Goualougo Triangle had a unique technological tradition just waiting to be discovered.

The Ndoki forest is never silent. You can always hear a cacophony of chattering, whistling, rustling, and crunching sounds as different creatures search for food, attempt to attract mates, or avoid predators. However, the rhythmic sound of a chimpanzee using a large branch to pound open a beehive is distinct from all the other noises in the forest. It sounds like someone chopping wood. The blows of the club reverberate in the high canopy, and the hollowness of the wood betrays that there is something hidden in the tree. This sound had frequently sent us racing through the forest in hope of catching a glimpse of a tool-using ape, usually to arrive only seconds too late to observe the ape in action.

Hearing it on what seemed to be a rather uneventful February morning in 2006, we ran along the path toward its origin. We arrived to find Dorothy lounging near a beehive surrounded by a swarm of stingless bees. She then picked up a large club that was lying on the branch beside her. She held it as if preparing to launch a javelin, but instead forcefully struck the hive entrance with the end of the club. She repeatedly pounded the hive for several minutes, then inspected the result of her efforts and took a short break. Much to our concern, she seemed willing to risk her life clinging, dangling, and lurching at various angles in the high canopy to breach the hive. She also showed us that several different types of tools may be needed to break into a beehive, which is sometimes lodged in a seemingly impenetrable structure of cemented mud or wood. Perhaps judging the first tool to be too large, she carefully stashed it in the canopy and broke off another branch, which she shortened and stripped of twigs and leaves to make a second, smaller club. She alternatively used the two tools as a hammer and lever to raid the hive.

It was strange to us that after spending nearly an hour opening the hive, she climbed into the leafy canopy of the tree without taking any honey, but we were quite impressed when she returned less than a minute later with a slender twig that was deftly fashioned into a dipping tool to extract the honey. Dorothy spent the rest of the afternoon enjoying the bounty of her tool-using skills. We wondered whether the other members of the Moto community might similarly take notice, or if they used the same techniques. Had Dorothy learned these skills in her previous community, or was this knowledge that Moto chimpanzees also possessed? Although this single observation answered several of our questions about tool use by chimpanzees in honey gathering, it seemed only a small preview of the tool-using traditions awaiting discovery in these forests.

Some days were better than others with regards to Dorothy's acceptance by the Moto group. We recall a misty morning when a party of mothers were foraging on tender leaf shoots near their night nests. It was a tranquil scene, where everyone seemed content to have found breakfast near to where they had slept the night before. We were startled when Dorothy suddenly became the victim of a seemingly unprovoked attack. She was quietly collecting the yellow-green leaves at the tips of a branch when an older female named Maya rushed toward her. Dorothy did not passively accept the attack, as we had previously observed her doing, but seemed to stand her ground and fend off the blows. From the intensity of the screams, we were certain that the young female was being ripped limb from limb. The other chimpanzees seemed to be jeering the opponents and nothing could be heard beyond their cries.

We rushed around the base of the tree in different directions to get a view of the fight. Maya suddenly separated herself from the brawl, and we were surprised to see that both the females were intact with no visible wounds. For a few seconds, everyone was still as the two females stared at each other with their hair standing on end. Maya then moved down the branch and lifted her arm to reach toward a female ally. The two older females rushed toward Dorothy, who immediately realized that she was outnumbered and fled from the tree. The older females seemed satisfied with their victory and settled down to intensely groom each other. Within minutes, the group was feeding and socializing as if nothing had transpired. Dorothy eventually returned to the tree and spent the rest of the afternoon in proximity to her attacker. She had withstood the tests of initiation for that day, but certainly could not have felt warmly welcomed by the matriarchs of the community.

In contrast to Dorothy's cold reception, we found that the females were often more accepting of our presence than the males. The young female was subject to several attacks, but the most serious assault on us consisted of showers of sticks and branches from the forest canopy. With that said, some chimpanzees have very good aim, and they even managed to dent one of our video cameras. Working in the forest with western lowland gorillas, we also knew how to respond, or rather show a lack of response, when charged by an angry ape. Most chimpanzee threats are bluffs, but that does not come to mind at the time. This would be the case when we were quietly making our way through a thicket toward the contented grunts of chimpanzees feeding on fallen fruits. Rather than use machetes to whack through the forest, we had long ago adopted the use of small garden shears to cut through thick patches of vegetation. The shears produce minimal damage to the vegetation and little noise as we track and follow apes. Each clip of the shears seemed loud and obtrusive on this quiet morning. Nonetheless, our arrival seemed unannounced to the small gathering of chimpanzees huddled in a small clearing within the tangled forest. Our Macallan was calmly grooming Maya, but his mood changed upon seeing us. His lips compressed and his shoulders hunched. He stood upright and ran straight toward us. At the last minute, he swerved onto the path that we had created through the thicket. We could hear him beating on buttress roots and calling to others as he traveled through the forest. The females then dutifully gathered their infants and began to follow his lead. We assumed our position at the back of the line, as the slowest and most socially inept of the apes.

There were milestones in both our and Dorothy's acceptance into the group, such as the first time that we found ourselves within a party of chimpanzees who were circling around a large tree to pick up fallen fruits and extract their seeds. Individuals would walk right past us without a sideways glance or walk up to within a few meters to pick up a fruit. Most of the chimpanzees were willing to tolerate our presence, but it would require their full trust to be able to closely follow each detail of their daily lives. By observing Dorothy, we learned that such acceptance could be earned through patience and respect for the social conventions of the Moto community. We followed suit by carefully studying social interactions and watching how different individuals responded to particular situations. As our observation skills improved, we found it easier to detect when a particular chimpanzee might be arriving to join the party or if tension was mounting in the group.

Our efforts were amply rewarded when we were searching for the chimpanzees in the heart of the Moto community range on a hot afternoon in February 2006. We caught a glimpse of Dorothy in the low canopy. She was turned away and hunched over. Sensing that this was not a typical situation, we slowly approached and waited for her to emerge from the thicket. A few seconds later, she climbed into the open with a tiny newborn clinging to her chest. This newest member of the social group was Dorothy's first contribution to the longevity of the Moto community and yet another hope for the preservation of these apes and their rich chimpanzee traditions.

FURTHER READING

McGrew, W. 1992. *Chimpanzee material culture: Implications for human evolution.* Cambridge: Cambridge University Press.

Sanz, C., and D. Morgan. 2007. Chimpanzee tool technology in the Goualougo Triangle, Republic of Congo. *Journal of Human Evolution* 52:420–33.

———. 2009. Flexible and persistent tool-using strategies in honey gathering by wild chimpanzees. *International Journal of Primatology* 30(3): 411–27.

Sanz, C., D. Morgan, and S. Gulick. 2004. New insights into chimpanzees, tools and termites from the Congo Basin. *American Naturalist* 164(5): 567–81.

Whiten, A., J. Goodall, W. McGrew, T. Nishida, V. Reynolds, Y. Sugiyama, C. Tutin, R. Wrangham, and C. Boesch. 1999. Cultures in chimpanzees. *Nature* 399:682–85.

SEVEN

Keeping It in the Family

Tribal Warfare between Chimpanzee Communities

JOSEPHINE HEAD

LOCATION: "Ozouga" ape research camp, 600 meters from the Atlantic Coast, Loango National Park, Gabon

August 10, 2005

I can hear the waves pounding against the beach in the background, and it lulls me into a reflective silence as I go through the day's events in my head, wondering how, despite twenty kilometers of patient walking through the woods, I failed to turn up any apes today. It is just before dusk, and I have returned to our coastal base camp after a day spent scouring the forest without success for signs of the elusive chimpanzees that we have been attempting to habituate for the past six months. The sun is setting over the ocean to the west, disappearing behind the tree line; soon we shall be left in darkness until tomorrow morning. A group of forest buffalo and a large solitary male elephant are grazing about 600 meters away on the narrow strip of savanna where our camp is located. I can hear our resident troop of red-capped mangabey monkeys bickering noisily just behind the palm-frond wall of our bathing-hut bucket shower. The heat haze is finally fading away on the horizon, and the gentle sea breeze whispers its way through the long dry grasses of the savanna.

Habituation of chimpanzees is a long process, which can take five or more years, but if you patiently make contact with the chimpanzees as often as you can find them, they will slowly but surely allow you into their lives, revealing the secrets of their forest world to you. Since childhood I have pored over books written by primatologists about chimpanzees and hoped one day to have the chance to study them in the wild myself. But for now habituated chimpanzees are still just a dream, and I must content myself with the odd glimpse of a retreating rump, since most days the chimpanzees are primarily concerned with getting away from this strange two-legged imposter as quickly as possible; sometimes stopping for a curious glance before vanishing back into the shadows.

Taking off my wet, muddy boots and beginning to mull over the pressing issue of whether to eat tinned sardines or corned beef for dinner, my silent reverie is rudely interrupted by a burst of chimpanzee calls just 400 meters south of our camp, and responding calls from 200 meters to the north. I leap up in excitement—since setting up camp here in February we have rarely heard chimpanzee vocalizations so close by. Our preliminary observations have led us to believe that our camp lies on the boundary between two chimpanzee communities, and we have always assumed that this zone is a kind of no-man's-land that both groups steer clear of to avoid conflict. I don't have time to put my shoes back on, so I just grab my binoculars and set off at a run in the direction of the closer calls from the north, the nearby buffalo swinging their heads up in astonishment at this sudden burst of activity so late in the day. I immediately surprise three adult chimpanzees who are about to exit the forest on the left, just in front of camp. They clearly want to cross the narrow strip of savanna in an attempt to join the second party of chimpanzees who are in the forest to the south. However, they hadn't counted upon my sudden arrival, and on seeing me so close by they immediately turn back and disappear into the forest, all too quickly becoming shadows and then melting into the leaf litter as if by magic.

Experience has taught me that I have little chance of catching up with them now that they know I am close by.

I rush back to camp and deliberate with the team about whether we should try and make contact with the second party of chimpanzees, which is now calling nonstop from the south. The "pant-hoot" vocalizations have been going on for some fifteen minutes, and there is clearly a lot of excitement. Pant-hooting is a chimpanzee-specific vocalization used for communication between individuals that involves breathing in and out repeatedly while making a hooting sound that builds into a crescendo and often ends in a near scream. Given that the general reaction from the chimpanzees to our presence at this early stage of the habituation process is to flee, we decide to leave them to make their nests near camp, and plan to get up early the following morning to make contact before they begin foraging for the day. This will give us at least the hope of being able to follow them for a little way, and of improving on our scant knowledge of their home range. It takes an iron will not to go running off after the chimpanzees when we know we risk missing them in the morning. We have to grab every chance that the chimpanzees decide to give us, and the sound of a group vocalizing, when you have spent days seeing and hearing nothing, is like being drawn to an addictive drug. But we all agree that this is the best plan, and we resist the urge to go running off into the dark toward the tempting calls. The excited pant-hoot vocalizations continue until 8 P.M. that night, which is unusual, since chimpanzees usually settle down to sleep by about 6:30 P.M. and only call at night if they are disturbed by something on the ground like an elephant or a leopard.

We spend the evening huddled around our kerosene lamp discussing what might have caused all this excitement. In the end, we decide that perhaps two parties of chimpanzees from the same community have encountered one another after a long separation and are expressing their excitement at being together again. Clearly, the boundary between the communities lies further to the south than we had previously imagined.

None of us have much experience of wild chimpanzees; after only six months working on this project, we are still learning many new things about their behavior every day. We go to bed early, already anticipating the thrilling contact we hope to make tomorrow with this large party of chimpanzees. The sporadic chimpanzee pant-hoots continue deep into the night, creeping into my dreams and creating jumbled visions of me running endlessly after vocalizations that never seem to get any closer.

August 11, 2005

We wake up in the dark and stumble around trying to feed and clothe ourselves as rapidly as possible in order to get a head start on the early-rising chimps. Slapping at mosquitoes as we wolf down leftover rice and sardines, we are surprised to hear so early a short burst of pant-hoots coming from the same location as last night. Quickly stuffing binoculars, GPS, and notebooks into bags, we run off toward them, desperate to contact the chimpanzees before they disappear deep into the forest for the day, where we might never catch up with them.

As we walk quietly and slowly approach the area where we think the chimpanzees nested, the first rays of daylight are beginning to penetrate the thick canopy, but we cannot see or hear any signs indicating that they still remain in the vicinity. Slightly deflated, we begin to scour the area for feces to be sieved as part of a study of the dietary habits of the chimpanzees, in addition to collecting genetic samples to try and estimate the number of chimpanzees in the area. We find eleven chimpanzee nests, all constructed the night before in the trees around a small stream, and set about collecting the feces, while also looking for any fruit remains that might tell us what the chimps have been eating. Additionally, we try to deduce which direction they might have taken this morning. Our searching leads us to a spot some fifty meters from the nest site where we discover an area of disturbed ground, with several broken saplings and shrubs, and more excitingly some tufts of black fur and what appear to be small chunks of flesh. It looks as though there was some kind of fracas here. The leaf litter is whipped up and some small saplings have been uprooted. Our excitement grows as we hypothesize that the chimpanzees must have hunted and caught a monkey yesterday evening, something that has been documented at several other sites where chimpanzees are studied, but that we have yet to confirm in our community of chimpanzees. Chimpanzees hunt cooperatively and even share the meat from the kill, but an absence in this area of their favorite prey, the colobus monkey, had led us to believe that perhaps they wouldn't do so here in Loango. Further investigation uncovers more tufts of hair, and then a trail of intestines leading behind a large tree. Hardly daring to hope that we could have been this lucky so early on in our study, we advance quickly expecting to find the remains of one of the five species of monkey present in the park.

I stop dead at the sight that meets my eyes, not really understanding what I am seeing and totally unprepared for it. A dead adult male chimpanzee lies spread-eagled on his back, his chest and throat ripped open, and his entire face and body covered with dozens of cuts and bruises. Much like someone faced with the aftermath of a gruesome car crash, I am at once unable to look at him, yet at the same time powerless to draw my eyes away. I feel as though I am looking at a person who has been murdered in a savage attack. Before we have time to digest this discovery, we hear chimps calling some 100 meters away and heading toward us, and we immediately retreat and hide 25 meters away behind a large tree, waiting to see what the chimps will do when they arrive. For rare situations like this, it is better to hide and observe the chimpanzees than to focus only on habituation, since at this early stage, our presence would clearly scare them away.

A party of nine adult chimpanzees arrives at the scene, and one large male makes his way directly to the corpse, sitting down beside it. Vegetation obscures our view and we are unable to see if he is looking at or touching the body. Two more males approach the body but then they spot us and immediately turn away and run off. This alerts the first male, who also runs away, closely followed by the rest of the party. When all the chimpanzees have left, we approach again, and study the dead male in more detail. The nature of his injuries and a lack of any other supporting evidence exclude a leopard as the perpetrator of this attack, and it is with a sinking feeling that we realize the only thing capable of inflicting such injuries is another chimpanzee. His spread-eagled position and the lack of any wounds on his back suggest that he was pinned down by several individuals, while the others attacked every available part of his body. Closer inspection reveals that his penis and testicles have been ripped off in the attack, and we presently find them some thirty meters away on the ground. We later learn that this is apparently a common trait in lethal aggression among chimpanzees, the emasculation of the offending male playing an unknown but important role in the ritual. The excited calls of last night suddenly sound like bloodthirsty cries of victory, and it is impossible for me to reconcile the "high-spirited" sounds we heard with this vicious attack. I am only too aware that as a field scientist, I am expected to be objective and never to attach human values to animal observations, but it is sometimes difficult to draw the line between the two when your study subjects are so similar genetically, and often behaviorally.

We return to camp that evening subdued and reflective, appalled and shocked by what we have seen. It is many weeks before the local field assistants can summon up any kind of enthusiasm for observing what they now consider to be the "savage" chimpanzees.

• • •

I reflected on the incident for days after the event, unanswerable questions rolling around in disarray inside my head. How could chimpanzees be capable of killing one of their own in cold blood? And how could they sound so carefree and joyful about it, not appearing to feel any remorse at their actions? Why did they come back and visit the body again the following day? Did they feel some kind of guilt, or just a detached interest in the corpse? Genetic sampling from the dead male and from feces found in the area that day confirmed it was the attacking chimpanzees that had come back to visit the body, rather than members of the dead male's own community come to investigate what had happened the night before. Intensive searching in the surrounding area after the incident showed that the attacking chimpanzees had come down from the north, and then quickly returned in that direction. We followed their trail for over a kilometer, feeling like detectives searching for clues at a crime scene. Footprints in sandy riverbeds and fecal remains on trails indicated their direction, and an absence of feeding signs showed that they had not stopped to feed at all along the way, leading us to believe that the attackers were from a different community to the victim. We had never provisioned these chimpanzees with food and they were still unhabituated to humans, and so we could only conclude that at least for Loango chimpanzees, lethal aggression is natural, and not something that occurs as a result of human interference.

The whole incident was so human in so many ways: physically, chimpanzees resemble humans, so the idea of several individuals pinning down one chimpanzee while they all took turns hitting and biting . . . it doesn't take too much imagination to see the parallels between this and gang violence where one person is brutally attacked by a large gang of overexcited, anger-fueled men. Yet at the same time the savagery and violence seemed somehow irreconcilable with human nature and what I have always wanted to believe is our innate disposition to be "humane." I am unsure if I was struggling with the shock of seeing firsthand the reality of chimpanzee aggression, which no amount of scientific literature can really prepare you for; or if I struggled more with the realization of how human the whole event was.

I think, in hindsight, that it was a mixture of the two. I was sickened by the action of the chimpanzees, but even more so by what it suggested about our own human nature. Is our "moral code" nothing more than a controlling system that humans have invented to keep some order in society? Are all human and animal societies bound together by nothing more than the selfish needs of each individual to survive? Is human morality really inherent or simply drummed into us from birth? Young children are supposed to have little idea of the difference between

INTERCOMMUNITY AGGRESSION

Chimpanzee communities throughout Africa are made up of many adult males living with many adult females and their offspring. Chimpanzees are highly territorial and actively defend their large home ranges, aggressively expelling any male intruder, but usually accepting female newcomers into a community. Territories are defended to protect food resources, females, and infants from neighboring chimpanzee communities. Intercommunity aggression in the form of attacks on chimpanzees from a different community is usually carried out by groups of adult males, which go on "patrols" into adjacent territories in search of vulnerable chimpanzees from neighboring groups who might be feeding alone or in smaller groups. They walk silently and purposefully, rarely feeding, and make as little noise as possible. If the patrolling males find an unknown chimpanzee in a tree, they will often try to surround it, blocking its escape before beginning the attack. Sometimes the victim is able to flee, particularly if other members of his community come to the rescue, but at other times the attack ends in death.

Intercommunity attacks have been witnessed in multiple chimpanzee research sites across Africa and have led to much discussion about the possible parallels between this aggressive behavior and the evolution of human warfare. In nearly all human societies studied to date there are extensive histories of warring between different clans or tribes, often involving the "stealing" of women and the killing of adult men in neighboring groups. Examining intercommunity aggression in chimpanzees is therefore useful for our understanding of this phenomenon in humans.

Jane Goodall first documented the killing of adult chimpanzees by other chimpanzees during territorial clashes in Gombe National Park, Tanzania, in the late 1970s. The same thing was later observed in ▸

▸ Mahale National Park, Tanzania, and Kibale National Park, Uganda. However, lower levels of intergroup killing have been documented at other long-term field research sites like the Taï National Park in Côte d'Ivoire and the Budongo Forest in Uganda, raising questions about the factors leading to the evolution of intergroup violence.

In addition to the killing of adult males, infant chimpanzees have been observed to be killed by neighboring groups of males on an even more regular basis, and in most instances their mothers are spared (see chapter 8). This is reminiscent of what has been observed in lions and in langurs (small monkeys found in Asia), where outsider males kill newborn babies when taking over a group. These killings allow new harem holders to invest in their own offspring immediately and for a longer period of time, as opposed to caring for the offspring of another male. This approach may thus be seen as a successful reproductive strategy imposed by males on females. In harem-living gorillas, following the death of a dominant male, silverback males have been observed to kill infants as part of the strategy for becoming the new leader. It may seem odd that females stay with a male who has killed their offspring, but this can be viewed as a sign of the male's strength and ability to prevent other males from doing the same thing. Similarly, we might expect female chimpanzees to join the males who have killed their infants, but only a few have been observed to transfer between groups as a consequence of such killings. Intercommunity violence also serves to reduce the size and strength of neighboring communities and discourages neighbors from expanding their territories. At the same time, it increases the home range of the perpetrators and hence their access to food and females. However, debate continues on the precise function of the intergroup aggression in this species.

"right" and "wrong," but in teaching the next generation how to be human, we nonetheless condemn "wrong" behavior as "savage," "animalistic," or "Neanderthal," comfortably drawing a solid line between ourselves and the rest of the natural world.

Or perhaps our own and the chimpanzees' moral code only extends to group members, and not to outsiders, in a kind of "us and them" mentality. Subsequent genetic analysis of the victim's tissue combined with genetic sampling of feces from throughout the study area supported our theory that the dead male chimpanzee was from a different community than the attackers, making the lethal attack a case of intercommunity aggression.

• • •

June 8, 2006

It is mid-morning, and we are walking in the forest, stepping quietly on the dry, leafy floor to avoid making any unnecessary noises that might frighten off the chimpanzees that we suspect are nearby. The *Saccoglottis gabonensis* trees look like gnarled old giants standing in groups, stooping and leaning in haphazard formation and moaning their complaints in slow motion as the wind pulls at their branches. For the last hour we have been zigzagging back and forth in an easterly direction trying to follow the tell-tale tracks and feeding signs that we have encountered occasionally, our only indications that chimpanzees have passed this way this morning. A stray leaf that has been kicked up, or a plant bent down, all these signs help us confirm that we are traveling in the right direction. Discarded fruit peels or chewed leaves are also very useful indications that the chimpanzees have passed this way.

The forest has an open understory, and we can see some eighty meters ahead of us. I spot a blue duiker (a small African antelope the size of a large rabbit) flitting among the leaves up ahead, only its flicking white tail visible most of the time as it vacuums up stray fruits left by other animals. He senses us and bounds away, emitting a high-pitched nasal cry and flashing his tail as he goes. Fresh feces and some more faint tracks in the sandy soil 50 meters ahead indicate that our chimpanzees are still heading east. Encouraged by our success, we follow the tracks and feces for another 150 meters, pursuing what appears to be a large party of chimpanzees. Although the chimpanzees are still not habituated to our presence, after eighteen months, we have become much more expert at reading the forest floor and picking out the near-invisible signs that indicate that chimpanzees have passed that way, and as a result our success in locating them has greatly improved.

The tracks we are following are just inside the range of what we have now confirmed to be the boundary between two neighboring communities; our only error was to assume that the chimpanzees avoided this zone. In addition to the fatal attack on the adult male last August, on one occasion we heard pant-hoots and alarm calls but arrived too late and only found a heavy trail of blood and diarrhea leading away from an area of trampled vegetation. A week later, we found the dead and heavily decomposed body of an adult male chimpanzee about 200 meters from the site. It was impossible to draw any conclusions about how he had died, since the forest here tends to reclaim its dead within a matter of days. On another occasion, we witnessed two parties of chimpanzees displaying aggressively at one another and vocalizing loudly, but they immediately saw us and both parties ran away in opposite directions.

Several minutes later, we are still following the tracks when the silence is suddenly broken by excited pant-hoots about 200 meters ahead. We sprint up and down three steep ravines in an attempt to reach the chimpanzees before they stop vocalizing, tripping over roots and being hit by low branches as we charge through the forest. There appears to be no danger of the chimpanzees quieting down, as the calls continue nonstop. We stop 60 meters away, breathing heavily, and are at once able to distinguish a single repeated alarm call among all the excited pant-hoots. Recalling the vicious attack of last year and our current location in the boundary zone between the communities, we lie down and begin to wriggle forward on our bellies, determined to observe exactly what is going on without the chimpanzees discovering that we are there.

We stop about 30 meters away from the location of all the noise and poke our heads up above a conveniently placed fallen log, a great spot for a bit of spying. We can see three large adult males running back and forth on the ground underneath an *Irvingia gabonensis* tree, displaying dramatically, with their hair bristling. The alarm calls appear to be coming from directly above them in the canopy. There are also pant-hoots up in the tree, and we assume that a chimpanzee has been trapped there and is being attacked by others. From our hiding spot we are unable to see little more than shaking branches in the canopy, but the displaying and vocalizing continues for seven more minutes before three more adult males descend from the tree, one of them displaying as he does so. He drags a heavy branch along the ground behind him, and then builds into his finale by rapidly drumming with his feet on the buttress roots of a nearby tree.

The last chimpanzee to descend the tree is an adult female, and it appears as though she was the one under attack by the males. She is still making repeated alarm calls as she reaches the ground and sits down, fearfully watching as the males begin to walk off to the north. Relieved that she looks to be in good shape and not badly injured, we watch as she follows after the males for twenty meters. But then she stops and picks something up, before immediately turning around and heading back in the opposite direction to the south, presumably wanting to increase the distance between herself and the other chimpanzees. We hear two more bursts of pant-hoots from the north and then silence.

Once the female is out of sight, we advance to the site where the incident took place, and immediately see that the ground beneath the tree is covered with many spots of blood and lots of diarrhea, a sign of fear in chimpanzees. We also find two small pieces of flesh covered in black hair, all adding to the evidence that the party of males had trapped and attacked the adult female before letting her go. I bend down to begin collecting fecal samples, and spot something pink poking out from under the leaf litter. Flicking back the leaves, I immediately recoil as I identify a very small and perfectly formed chimpanzee foot lying on the ground, severed at the ankle. A lump rises in my throat and I think I might vomit as I realize with a sinking heart that it wasn't the female who was under attack . . . it was her infant. I become conscious that the small bundle we saw her picking up before leaving was almost certainly her infant, and I remember it was at this point that she finally stopped making alarm calls, having finally been reunited with her baby.

We are less than 500 meters from the site of last year's fatal attack on the adult male. Recalling the duration and intensity of the assault combined with the obvious severity of the injuries, it seemed very unlikely that the infant survived the attack, and that the males had simply lost interest and departed once they had killed it. All the same feelings of disbelief and horror come flooding back in an instant, and somehow last year's fatal attack on the adult male has in no way prepared us for this new incident. The chimpanzees that we are finally beginning to identify, and whose different personalities we are so eager to get to know, are yet again exhibiting violent tendencies that on the surface seem difficult to justify.

• • •

Thinking about things rationally instead of emotionally is normally a better plan when considering the reasons behind what might otherwise appear to be senseless and cruel behaviors. Although it may seem incomprehensibly unnecessary to kill a defenseless infant, there are clear and sometimes immediate benefits to the attacking chimpanzees in doing so. Not only are they weakening the potential strength of the neighboring community by reducing its numbers and thus protecting themselves against future attack, they are also encouraging the bereaved female to join their community, since succeeding in killing her infant proves to the female that her own community is not strong enough to protect her and her offspring. Essentially the killing of her infant can actually encourage a female to join and mate with the perpetrators. Killing off the adult males in a neighboring community serves a similar purpose, reducing the strength of the rival group while increasing access to females and food resources. A female chimpanzee in a community with few remaining males is likely to transfer to a group that has many adult males, increasing the protection for herself and her future offspring.

Human definitions of "right" and "wrong" are just that: human; and they have evolved to make modern human society function positively. But that does not mean that our own social rules should be the benchmark for all other animal societies, since each species has evolved its own "moral code" that has adapted as a function of their own particular environment. Within a community, chimpanzees can be very caring, licking and cleaning the wounds of other individuals over a long period of time, and patiently waiting for injured chimpanzees to catch up with the rest of the party. So perhaps we should not be so quick to condemn the chimpanzees for their behavior, and should instead look at their actions from a different perspective, one that can teach us more about ourselves than perhaps we would like to admit. Perhaps we humans are not so very different from our chimpanzee neighbors, but have just learned to control our aggression in certain situations, and perhaps our different social models have resulted in different moral codes. In a world where just about everything human and animal seems to be guided by an inherent need to thrive, the chimpanzees are simply trying to assure the successful future of their community in the same way that humans did and still do: by protecting themselves.

FURTHER READING

Boesch, Christophe. 2009. *The real chimpanzee: Sex strategies in the forest.* Cambridge: Cambridge University Press.

Boesch, C., J. Head, N. Tagg, M. Arandjelovic, L. Vigilant, and M. M. Robbins. 2007. Fatal chimpanzee attack in Loango National Park, Gabon: Observational and genetic evidence. *International Journal of Primatology* 28:1025–34.

Goodall, Jane. 1986. *The chimpanzees of Gombe: Patterns of behavior.* Cambridge, MA: Havard University Press, Belknap Press.

Watts, D., M. Muller, S. Amsler, G. Mbabazi, and J. Mitani. 2006. Lethal intergroup aggression by chimpanzees in the Kibale National Park, Uganda. *American Journal of Primatology* 68:161–80.

Wilson, M.L., W.R. Wallauer, and A.E. Pusey. 2004. Intergroup violence in chimpanzees: New cases from Gombe National Park, Tanzania. *International Journal of Primatology* 25:523–50.

Wilson, M.L., and R.W. Wrangham. 2003. Intergroup relations in chimpanzees. *Annual Review of Anthropology* 32:363–92.

EIGHT

Winona's Search for the Right Silverback

Insights into Female Strategies at a Natural Rain Forest Clearing in Northern Congo

THOMAS BREUER

August 2007

It was another rainy month in the Nouabalé-Ndoki National Park in the Republic of the Congo. My research assistants and I were sitting on the observation platform that overlooks the edge of the Mbeli Bai forest clearing. The heavy raindrops were a great relief, because the rain stops the innumerable sweat bees—small nonstinging flies that are attracted to the salts in human perspiration—that would otherwise get into our eyes, ears, and nostrils. They are normally an unavoidable irritation—there is nothing you can do about it.

It was one of those quiet days with no animals to watch in the forest clearing, but then we suddenly saw a reddish head popping out of the forest edge, around 200 meters from our observation platform. The female western lowland gorilla carefully and silently walked over a swampy patch of aquatic herbaceous vegetation. We were having difficulties identifying the gorilla with our spotting scopes, because the grass was very high and she had her back to us. After about two minutes, however, she turned around, and we could see her face. My assistants simultaneously said "twins" when they saw her holding two infants to her chest. I spontaneously added "Winona."

This is the story of Winona's search for the right silverback.

• • •

Winona is one of many western lowland gorillas that have been identified and named by researchers at Mbeli Bai since 1995. Mbeli Bai is a thirteen-hectare swampy clearing in the southwest of the Nouabalé-Ndoki National Park. The word *bai* means forest clearing in the local language, Lingala. Mbeli Bai is the largest natural forest clearing in the region, to which western gorillas come to feed on mineral-rich aquatic herbs that grow only in the swamp.

Working at a bai and waiting for gorillas to appear is very different from following a habituated group in the forest. Since it is known that many animals come into the bai regularly to feed, the idea is to wait and observe what appears, rather than go in search of the gorillas in the forest. This can involve waiting for hours or even days on the observation platform for gorillas to show up at the clearing; all we can do is watch and wait. Our aim is to disturb the gorillas as little as possible in the forest. We have only very limited knowledge about what the gorillas are doing and where they are ranging when they are not in the bai. It is impossible to predict who will appear next at the bai. In the evenings we often sit together at the camp and discuss the day's sightings, but we can only speculate about the animals that will appear next at the clearing. Fortunately, there are also forest elephants, buffalo, and sitatunga (marshbuck) to observe, making the days without any gorillas pass quickly. Needless to say, we also get very excited whenever gorillas do come to the bai.

The vegetation in Mbeli Bai is a floating mass about one meter deep over water-logged soil. If you walk into the bai, you quickly sink into the water, and if you step on the wrong clump of vegetation, you may slip breast-deep into the mud. Gorillas are well adapted to the swamp environment, however, and they often walk bipedally through the deep aquatic vegetation.

The naturally treeless clearing of Mbeli Bai offers a unique perspective from which to study gorillas in the rain forest. Western gorillas are extremely difficult to see in the dense vegetation of the forest, and because of their elusive nature, there are very few groups of them that are habituated to the presence of human observers (see chapter 9). Mbeli Bai is unusual because it overlaps with the home ranges of many gorilla groups, so we are able to observe many different groups that are attracted to it and learn a great deal about aspects of group composition and how individuals disperse. Sometimes more than one group is in the bai at the same time, making it possible to collect detailed behavior on their interactions.

Although the research at Mbeli Bai provides only snapshots of gorillas' lives, it allows us to observe many different western gorillas and gather information on their demography, important life history events, and birth and mortality patterns. Because we observe so many gorilla groups over the course of time, we are able to gain a population-wide perspective on these events much faster than through studies that follow only one or two habituated groups. Frustratingly, most of these life history events occur when the gorillas are not in the bai. Additionally, there are typically long gaps in time, spanning weeks to months, between visits made by any particular group. Sometimes we thus have to deduce what events have happened, such as whether a gorilla has died or simply migrated into a group that doesn't visit the bai. Over the past fifteen years, we have monitored more than 380 western gorillas. Currently, Mbeli Bai is visited annually by around 130 gorillas belonging to fourteen social groups, and ten solitary males also come there. When I took over as principal investigator in 2002, Winona's life had already been followed for over seven years by my predecessors.

• • •

In contrast to mountain gorilla groups, 40 percent of which contain more than one silverback, groups at Mbeli Bai, like almost all western gorillas, have only one silverback. Adult female gorillas always reside with a silverback, doubtless because they need the protection provided by an adult male against other, competing silverbacks. Females thus choose to stay with strong silverbacks that will protect them and their offspring against infanticide by other males. Female gorillas regularly transfer between different groups, and researchers distinguish between voluntary and involuntary transfers. In a voluntary transfer event, females transfer from stable, intact groups. However, if the sole silverback of a group dies, its remaining members have to find a new group or silverback—this is an involuntary transfer. Given that a silverback who is not the father of their infants might kill them, females with unweaned offspring have to make the important decision of which new silverback to choose.

Researchers have speculated if and how a female gorilla might avoid losing her unweaned infant to infanticide when a silverback dies. Winona has given us some clues as to how females might lower the risk of infanticide risk by choosing the right male when she is forced into an involuntary transfer. Hers is the first case indicating that the age of the new silverback that a female joins might be a key factor in lowering the risk of infanticide. On another level, the events of Winona's life make a fascinating story and exemplify the soap-opera aspect of Mbeli gorillas' lives.

• • •

Winona was one of the first gorillas identified at Mbeli Bai. In March 1995, she was observed in a group named "Clive," after its leading silverback. When first seen, Winona was one of two adult females in Clive's group, and she had a newborn daughter we called Wendy. There were also three other young females in the group, but we don't know if any of them were related to Winona.

In August 1998, Clive lost all three of the young females to an impressively large silverback, Khan, who was new in the Mbeli population, but Winona and the other adult female remained with him. It is possible that these young females were Clive's daughters; if so, it is

INFANTICIDE

In many mammalian species, particularly primates and social carnivores, males sometimes kill infants. From a human perspective, this is not a very nice side of animal behavior, leading to the question of what the logical explanation is from an evolutionary perspective. Why do males commit infanticide?

The most widely accepted explanation postulates that males gain a reproductive advantage by killing infants that they did not sire. This decreases the reproductive success of the male that did sire the offspring and in many cases increases the possibility of the killer fathering offspring with the mother of the infant that was killed. Females are constrained in the timing and number of offspring they produce by the long time necessary for gestation and lactation, which may be from three to five years or more in great apes. During this time, females are unable to conceive, which makes biological sense, since it would be undesirable for them to have more dependent offspring than they could successfully care for at one time. In contrast, males' ability to produce offspring is largely constrained by their ability to find partners. So if a male joins up with a female who has a newborn baby, he has the choice of either waiting until that infant is weaned, which will take years in gorillas or chimpanzees, or killing the infant, so that the female quickly becomes sexually receptive again and can produce an infant with him. Thus infanticide reduces the time a male has to wait to be able to mate with a particular female, reduces the reproductive success of other males, and potentially increases his own reproductive output.

The consequences of this reproductive strategy are wide-ranging. First, it means that males will want to protect offspring they have sired. To do this, they need to stay with the mother until the risk of infanticide has passed, which leads to longer-term associations between males and females. Conversely, females will want to reduce the risk of infanticide so it is in their best interest to garner protection from the father and stay with him. In fact, the risk of infanticide has been put forward ▸

▸ as one of the strongest forces leading to long-term associations or bonds in primates and some other mammals. It is at the root of why primates live in social groups.

Another implication of the risk of infanticide is that females should seek out mates who will be best able to protect their offspring from other males. Males that are bigger and stronger will be better at warding off other males and so are preferred by females. Over time, bigger males thus produce more offspring than smaller males, which leads to males typically being larger than females in most primate species. Adult male gorillas are nearly double the size of adult females. The size difference between males and females is not as great in chimpanzees and bonobos, but the males are still clearly larger than the females. Being bigger also has many other implications for how the sexes interact (see chapter 4).

You may now have the impression that male apes have the upper hand in things, but female apes have developed a number of counter-strategies to reduce the risk of infanticide besides forming long-term pair bonds with the father. The main approach is to "confuse paternity" by mating with many males, so that none can be certain of being the father or not. If a male has reason to suspect that he is the sire of an infant, he is likely to err on the safe side and not kill it. This strategy works well in species that live in groups containing several adult males, such as chimpanzees and bonobos, in which we don't see infanticide by other males in the group, but we do see infants being killed by males from neighboring communities.

What about gorillas that live primarily in one-male groups? Yes, their options are more limited. In the cases in which mountain gorilla groups contain more than one male, females typically mate with more than one male. However, western gorilla females may only mate with the one silverback in the group. But over a longer timescale, female gorillas show their choices in males by transferring between different groups. Obviously, females don't want to transfer when they are ▸

not surprising that they transferred to another group. In April 2000, the group increased in size when Clive gained two adult females and one young female. The young female, whom we named Misha, suffered a lot of aggression from the other group members, particularly from Winona. Since there does not normally seem to be much aggression between females over food—which is superabundant in the bai—this was unusual. Winona might have been concerned about her dominance rank in the group and wanted to demonstrate this clearly to the new female. It wasn't until early 2001, nearly six years after Wendy's birth, that Winona gave birth to her next offspring, a son we called Whisky. Whereas mountain gorillas give birth every four years, western lowland gorillas take around five and a half years, thereby extending the time period during which infants are susceptible to infanticide.

In mid 2001, Clive disappeared, although when we last saw him, he

▸ pregnant or nursing, so the window of time in which they can transfer may be only the six months or so, out of every four or five years, when they are sexually receptive. Some females may reside with one particular silverback for more than a decade, whereas other females may move to a new group after each of their offspring is weaned. The option of transferring is also dependent on how frequently different groups meet in the forest. If a silverback doesn't want the females of his group to leave him, he might try to avoid meeting other groups; we'd expect this to be the case for older, weaker males. On the other hand, a solitary male or a male trying to increase the number of females in his group may make more of an effort to interact with other groups. Presumably, females prefer males that are young, strong, and capable of being good protectors until an infant is weaned. However, in cases where a female has an unweaned infant when her silverback dies, she may be forced into making some difficult decisions about the best choice of group to prevent her infant from being killed.

appeared to be in good health. Winona, Whisky, Wendy, Misha, and two juveniles from Clive's group were seen together without any silverback on two subsequent visits into the bai, but Misha's newborn baby was never seen again. Although we have never witnessed infanticide in the bai, we frequently don't see young babies again after their mothers involuntarily transfer into new groups, which can serve as circumstantial evidence of infanticide, even though infants may have died of other causes. One of Clive's young females that had transferred to Khan's group was seen with the shredded remains of her baby on her back during the first visit to the bai after her transfer, providing more compelling evidence. We can only speculate, but if Khan had never copulated with this female, he probably knew that he wasn't the father and killed the baby, believing it to be Clive's. It is also possible that Misha's baby was killed by another male, since presumably other males were trying to obtain leadership of this silverbackless group.

We were worried about the fate of Whisky, given that he was less than a year old when Clive disappeared and extremely vulnerable to infanticide without the protection of his father. Influenced by the pictures and descriptions of Dian Fossey, gorillas have been known as "gentle giants." Certainly gorillas can be very gentle, and most of the time they are very peaceful. However, whether they are mountain or western gorillas, the interactions among individuals when two gorilla groups meet can be very brutal, and physical aggression is not uncommon. Strength and fighting ability are key features influencing an adult male's ability to attract and retain females. If a female transfers voluntarily, she might thus be expected to choose a younger, bigger, stronger silverback, who can ensure the survival of her offspring. Yet this is the type of male most unlikely to tolerate a young infant sired by another male. So Winona needed to find the opposite—a silverback who would accept her into his group without harming Whisky.

We were naturally anxious to see whether Winona would be able to find a male that would accept both her and Whisky. Her task was not an easy one. However, we have found over the years that old silverbacks at

Mbeli attract a lot of individuals that have difficulties joining another group. These old males seem to be very tolerant and not aggressive toward other gorillas, particularly juvenile and subadult males that are typically not tolerated by silverbacks that aren't their close relatives.

The subsequent choices Winona made of which groups to join reflected her attempts to avoid infanticidal males. More than four months after last being seen in the remnants of Clive's group, Winona, Whisky, and the two juveniles appeared in Mosombo's group. Mosombo was an old silverback who was first observed at the bai in 1995, together with his presumed older son and members of another group that had disintegrated. Winona knew Mosombo from various encounters in the bai, and perhaps she somehow knew that he was old. She had certainly made a good choice with Mosombo, because she was accepted into his group and he also accepted Whisky, who was only seven months old.

Unfortunately, Mosombo was old and weak, and his tenure lasted only for another six months before he died. So joining him had perhaps not been the best choice after all, since Winona now faced the same situation as when Clive died: the leading silverback of her group had died and she had dependent offspring who would not survive a transfer to a young aggressive silverback. Winona was seen in the bai on various occasions with other members of Mosombo's group, but with no silverback. It was four months before she finally joined what was called Noodles group, which had been known to us and to Winona since 1995.

The dominant silverback of Noodles group, Basil, was also old and two of his presumed sons had already reached maturity and emigrated from his group. Winona and Whisky were quickly accepted by him. Besides Winona, there was only one other adult female, named Salmonberry, who had joined the Noodles group two years before. She was quite famous at Mbeli for having given birth in her former group to a snow-white albino infant, who had unfortunately died. Now Salmonberry had a child with Basil, a daughter named Sage, who became a good play partner for Whisky. At this young age gorillas always make us laugh, because they love to scratch their bellies and stick out their tongues.

• • •

Almost all the twelve members of Noodles group were immigrants from other groups. Winona had made another good choice, since Basil was a very gentle old silverback. Noodles group members were by far the most frequent visitors to Mbeli Bai. In September and October 2002, they were visiting the clearing nearly on a daily basis, mainly to feed on a succulent fruit species called *Nauclea diderrichii* that tastes like pineapple. However, at the end of September Basil was seen with a massive wound on his right shoulder, and he looked very thin. Although we have never seen serious interactions between silverbacks in the bai, intense male-male aggression is probably a much more common feature in the forest than we suspect. In the open clearing, interactions are much more ritualized, and silverbacks mainly exchange displays such as chest-beating, strutting walks, or head-turning (in which they show off the large crests on their heads).

Throughout the following weeks, Noodles group encountered many other groups in the bai that were also attracted by the nauclea fruits. During these visits, Basil was very inactive and spent most of the time picking his wound. Other group members groomed Basil on a regular basis, particularly two of his presumed sons. Other silverbacks frequently approached Basil and displayed as close as one meter in front of him, but Basil was too weak to react to this intimidation. The older blackbacks, particularly his son Coriander, often intervened during these interactions by running between the two silverbacks. My team and I hoped that Basil would recover, because we liked his calm disposition toward the other gorillas in his group and we were worried about the future of Winona's and Salmonberry's kids. Unfortunately, however, Basil disappeared and presumably died. The young silverback Coriander replaced him as head of Noodles group for a couple of months, but he was still young and inexperienced. During Coriander's short tenure as leader, he also did not kill Whisky, who thus managed to survive the death of another silverback.

In the following three months, nine other gorillas immigrated into Noodles group, including an old silverback called Atticus, increasing the group size to eighteen. Atticus was able to dominate Coriander and became the leader of the group. Coriander became increasingly peripheral and finally left the group to become a solitary silverback. Atticus was very aggressive toward Winona and often stayed close to her throughout the group's visit to the bai. Fortunately, Whisky was now over three and a half years, and since he was almost weaned, he was much less vulnerable to infanticide.

Atticus's reign proved to be short-lived. A few months later, a young silverback, Voldy, immigrated into the group, and Atticus was observed with large wounds. Within a few months of Voldy being in the group, Atticus disappeared. Noodles group was without a mature silverback, and again we worried about the future of Winona and Whisky. At this point in May 2005, Whisky was five years old and fully weaned, so infanticide was no longer a risk, but we were still concerned about whether he would be able to grow up in a stable group or not.

Half a year later, presumably because she didn't view Voldy as mature enough to be a strong leader, Winona left Noodles group and joined a solitary male named New Vidal for a few months, but did not have a child with him. Winona left Whisky behind in Noodles group, which is commonly done by females when they have weaned their offspring. She was not seen for some months and visited Mbeli Bai again in August 2007 with another silverback—one we had known for a long time and one long familiar to Winona too. It was Coriander, the male she had lived with for some time in the Noodles group. But perhaps the biggest surprise of all was seeing Winona with twins, something that has been observed only rarely in gorillas. In 2009, the twins were two years old and Winona was one of five females in Coriander's group. The twins appeared healthy and Winona was proving a good mother with the challenge of two infants instead of one.

• • •

Winona's story is an example of the dynamic society of western gorillas. In total, Winona lived with six different silverbacks over a fourteen-year period. Her story exemplifies how the small changes we witness on a daily basis play out in an individual's life. There is a story to tell for the hundreds of gorillas we have observed in the bai, even if we can't observe them all of the time. Typically, if a female transfers from a habituated group to an unhabituated one, researchers have no idea what happens to them. Only through being able to observe her in a population that regularly visits a bai was it possible to follow her life. She also added important information to our understanding of the strategies females use in choosing a mate. By transferring several times to older males that wouldn't usually have been considered good options, she showed that mate choice is influenced by many different factors and is actually very complex.

In contrast to many other regions in Central Africa, to our knowledge, gorillas have never been hunted at Mbeli. Mbeli Bai was a major elephant poaching area before the monitoring began in the 1990s, but since then no elephants have been killed. The Wildlife Conservation Society, recognizing the exceptional wildlife and unique aspects of the forest, helped to create the national park together with the Congolese government. Mbeli Bai is one of the last remaining pristine habitats in Central Africa and attracts more and more visitors from all over the world with whom we like to share our experiences. Stories like Winona's are always special to both researchers and tourists visiting Mbeli Bai. Telling Winona's story to local children is also always a great pleasure, because they find it so fascinating. We also run a conservation education program called Club Ebobo (*ebobo* is the local word for "gorilla"). Currently, around 1,000 schoolchildren, many of them seminomadic pygmies, attend our monthly lessons on a frequent basis. Until now, none of the children knew that every gorilla has its own unique story. Needless to say, it is important to take the stories hidden in the forest to the people of Congo and beyond.

FURTHER READING

Breuer, T., A.M. Robbins, C. Olejniczak, R.J. Parnell, E.J. Stokes, and M.M. Robbins. 2010. Variance in the male reproductive success of western gorillas: Acquiring females is just the beginning. *Behavioural Ecology and Sociobiology* 64:515–28.

Breuer, T., M.M. Robbins, and C. Boesch. 2007. Using photogrammetry and color scoring to assess sexual dimorphism in wild western gorillas (*Gorilla gorilla*). *American Journal of Physical Anthropology* 134:369–82.

Harcourt, A.H., and K.J. Stewart. 2007. *Gorilla society: Conflict, compromise, and cooperation between the sexes.* Chicago: University of Chicago Press.

Levréro, F., S. Gatti, N. Ménard, E. Petit, D. Caillaud, and A. Gautier-Hion. 2006. Living in nonbreeding groups: An alternative strategy for maturing males. *American Journal of Primatology* 68:275–91.

Stokes, E.J., R.J. Parnell, and C. Olejniczak. 2003. Female dispersal and reproductive success in wild western lowland gorillas (*Gorilla gorilla gorilla*). *Behavioural Ecology and Sociobiology* 54:329–39.

Watts, D.P. 1989. Infanticide in mountain gorillas: New cases and a reconsideration of the evidence. *Ethology* 81:1–18.

NINE

The Long Road to Habituation

A Window into the Lives of Gorillas

CHLOÉ CIPOLLETTA

March 14, 1998

Another long day has passed. We are on our way back to camp, tired and frustrated. For the fifth day in succession, we have been searching for gorillas since early in the morning, but have only come across a few old feeding remains and found no fresh signs that may lead us to the gorillas. I begin to wonder about the task I have recently committed to, when my thoughts are abruptly interrupted. A terrifying scream echoes from behind our shoulders. We have just walked past a gorilla family! We missed their trail and did not see a leaf move: thankfully, the silverback had something to say. While his screams are meant to scare us away, we welcome this sign of life. We mark the place with a few broken twigs and head back to camp before darkness. Tomorrow we have the whole day to catch up with the group, following the traces they will leave behind as they eat and move across the forest. With some luck we shall be able to make another contact.

• • •

Early in 1998, I left the noisy streets of Rome for Dzanga-Sangha in the Central African Republic (CAR), on the northern edge of the

Congo Basin forest. Supported by the World Wildlife Fund for Nature (WWF), I was to lead a program to develop tourism based on the tracking and viewing of western lowland gorillas, a strategy to generate much-needed revenue for both the local economy and the ongoing conservation efforts at Dzanga-Sangha. The first step involved habituating a group of gorillas, but I had had no previous experience with them. I had, however, just spent two years in Côte d'Ivoire, West Africa, where I learned the techniques to habituate chimpanzees. The first step to habituation is simply finding the apes. With chimpanzees I used to spend hours in the forest waiting to hear them hooting or drumming: the same technique chimpanzees use to find other members of the group (or to avoid them). Drumming is a particularly powerful way of displaying and communicating: chimpanzees will choose trees with large buttress roots and energetically beat them with their hands and feet, producing an impressive drumming sound. Because this drumming sound can carry over a kilometer in the forest, it's a signal we found very useful to locate the chimpanzees. Gorillas, however, don't drum or regularly make loud hooting vocalizations, so other methods of finding them are necessary.

Several factors help researchers find mountain gorillas in the forests of Uganda, Rwanda, and the Democratic Republic of the Congo (DRC). First, since mountain gorillas live in what amounts to a "salad bowl," they don't need to go far to forage; second, they leave behind a multitude of feeding remains, precious clues for anyone trying to locate them; and, third, the ground is covered by soft herbaceous plants that bend when walked on, pointing straight in the direction the gorillas have taken.

In contrast, western lowland gorillas are rather elusive, and finding them can be tricky, because there is so little herbaceous vegetation in their habitat. The forest floor, covered by dead leaves, roots, and rotten vegetation, is a perfect carpet that hides knuckle prints and other signs of gorilla movements. To fulfill their nutritional requirements, every day western gorillas cover over twice the distance walked by mountain gorillas. The main reason for this difference is fruit: western lowland gorillas roam through the forest in search of ripe fruits, which are sparse and scattered. With the lower availability of edible herbaceous vegetation, daily feeding trails are longer and more difficult to piece together in the case of lowland gorillas than for mountain gorillas. To make things more difficult, even when in the vicinity of a gorilla in a lowland forest, it would be rare luck actually to see something more than a dark shadow and some movements in the vegetation. Visibility is very limited by the thick tangle of vines and shrubs that link the trees together, constituting what is known as a dense tropical forest.

Moreover, western gorillas have suffered much more from illegal hunting than mountain gorillas, making western gorillas even more wary of humans. It is thus no coincidence that western lowland gorillas, the most numerous of the gorilla subspecies, are still little known. Habituating a group at Dzanga-Sangha would not only enable the tourism program to succeed, it would at the same time allow us to gain more knowledge of these close relatives of ours.

From the beginning of our program I was lucky enough to work with skilled Ba'Aka "pygmy" trackers, local people with an intimate knowledge of the forest and its inhabitants. What to an untrained eye is a simple twig, bent like many others, is pieced together by the trackers with other imperceptible clues, a smell, or the wet surface of a dead leaf turned upward, and a gorilla trail emerges apparently out of nowhere. Anyone who has had the chance to observe a Ba'Aka track gorillas in the forest would admit there is some kind of magic to it. It is the ancient knowledge of hunters and gatherers, put into practice for something many of them view as complete madness: the habituation of gorillas!

The search for gorillas is a challenge that Ba'Aka hunters are ready to take on. Traditionally, they would occasionally silently sneak up to a resting gorilla and throw their spear before the gorilla could react. It is a very dangerous way to hunt, as testified to by a few elders in the villages who bore the scars of gorillas' fury. Indeed, most Ba'Aka are scared of

WHAT IS HABITUATION?

The process through which wild animals learn to accept the presence of human observers in their natural environment is called "habituation." It is not only the first step undertaken by most researchers wishing to study primate behavior in the wild but also the preliminary work needed to develop tourism that involves ape viewing. To habituate a gorilla family, the observers must engage with the same individuals in a series of repeated peaceful "contacts." Thus, the key to habituation is being able to consistently locate the same group on a regular basis. During these contacts, to enable the gorillas to quickly recognize and distinguish them, it helps if the human participants behave as consistently as possible with respect to the sounds they make and their appearance. Typically, gorillas display a gradual change of reactions as they go through the habituation process. Their initial fearful responses, running away and screaming, next turn into more serious aggressive threats, with loud vocalizations and direct charges led by the silverback. This is the most challenging phase, since the observers must hold their ground without appearing to challenge the leader's authority. Gaining the trust of the silverback, the leader of the group, is a crucial step, because his reactions will influence those of the whole group. Eventually, the aggressive reactions will give way to curiosity: once humans are no longer perceived as a danger, they become the object of curiosity, especially among the group's juveniles. Habituation is achieved when the observers are allowed to walk behind the group members without distracting them from their daily activities.

A series of factors may influence the speed of habituation and determine whether it will be successful or not. Among these, the most important is the ability to establish regular contact with the same group, which in turn depends on the environment where the animals live, their ecology, and how many groups live in a particular area. Another factor thought to influence the likelihood of success is the nature of the animals' past experience with humans, particularly if there is a history of hunting in the area where habituation is taking place. Finally, individual ▸

▸ characters also play an important role, because some group members may never be fully at ease around humans, while others will be so comfortable as to fall asleep only a short distance away from someone.

Whether it is for research or tourism, the habituation of wild animals involves some important risks, which must be addressed beforehand and considered throughout the process. Regular close encounters between human and wild animals increase the likelihood of disease transmission. This concern is particularly relevant for all great apes, whose close relatedness to humans makes them susceptible to many diseases for which they have no immunity, lacking previous exposure to them. Outbreaks of respiratory disease have occurred in several groups of habituated gorillas and chimpanzees across Africa, with many animals dying. To minimize the risk of disease transmission, strict guidelines must be developed and enforced to guarantee that no sick person comes in close contact with wild apes and that a minimum recommended distance (generally between seven and ten meters) is kept between the visitors and the animals, and to define adequate modes of conduct (i.e., covering one's mouth when sneezing or coughing, wearing surgical masks, properly burying waste, etc.).

Another health concern is that the gorillas may be more vulnerable to illnesses simply because they experience more stress because of the presence of humans. If humans somehow modify the habitat and hence food availability, or alter the behavior of the gorillas (e.g., if the gorillas are less likely to feed while observers are present), they may also be more susceptible to disease. These concerns are best addressed by limiting the time and total number of people visiting a group each day, as well as by ensuring that facilities are built with a minimal impact on the environment. The overall presence and impact of people in the area where animals are habituated should be kept to a minimum.

Finally, projects involving habituation of wild animals must make a long-term commitment. Since habituated apes are vulnerable to hunting once they have lost their fear of humans, continuous protection of their habitat is required to guarantee their safety.

gorillas and would not recommend getting close to a living specimen. For our work, we needed to silently approach a gorilla family and, from a distance, let them know we were nearby. We did this by making a loud clack with our tongue, a signal we would repeat at every encounter and which the gorillas would gradually associate with us. Over time we would work at decreasing our distance to them. Indeed, to many of the trackers, letting the gorillas know about our presence seemed to just spoil the hard work put into finding them: within a second of hearing us, the whole group would disappear, running away from the terrifying humans.

One cannot blame the gorillas for fleeing: illegal hunting is a plague in central African forests. In 1991, the government of the CAR and the WWF joined forces with the objective of protecting the largest remaining tract of forest in the country. While poaching is still an issue (in 2004, nearly 50,000 snares were taken out of the forest, and 54 guns and 689 rounds of ammunition were seized, together with 1,400 kilograms of bushmeat and fifteen elephant tusks), the wild populations still found in the area, notably gorillas and elephants, are living testimony to the vital importance of the conservation efforts carried out over the years. Without the continued struggle to stop poaching, the park would soon be empty of these amazing creatures.

• • •

September 7, 1998

We have been following the trail of a gorilla group for several months now, and our encounters are becoming more regular. I am happy about the progress achieved so far: being able to establish regular contact with the same group was our first challenge in the habituation process, and we achieved it. Today I am determined to fight the Ba'Aka's reluctance to give gorillas, or any animals, names. I figure the time is right. The group we are following has already started acting differently from the others, proving my point against theirs: not all animals are the same! The gorillas help me in the task. As we move from our last place of contact, assuming the group had moved ahead to avoid us, we are surprised to find the silverback and two juveniles sleeping some ten meters away in the thick vegetation. Instead of screaming, charging, or running away, they give us a perplexed and annoyed look and slowly walk away only to stop after a few meters to resume their rest. Indeed, after months of meeting us in the forest, hearing us clacking our tongues and never attempting to hurt them, the gorillas are learning that while we may be annoying and disturb their naps, we are not dangerous. Thrilled, I look at the trackers and ask: what do you think? "Group Munye!" exclaims Molongo: the "good" group finally found a name.

• • •

It is hard to say how many individuals were in the Munye group when we first met them. Given the low visibility in the forest and the gorillas' fearful responses to our approaches, all we saw in the first months were dark shadows. We did get to know Mlima, the silverback, quite early on. He took his role as the group's protector quite seriously and would generally be the first to "greet" our arrival with loud and intimidating screams. Even when he seemed to start accepting our presence, Mlima would not miss the chance to show off in front of his females. He would exhibit impressive displays toward us when a female was looking, but if no female was nearby, he'd calmly sit not too far from us feeding or resting.

Almost daily we would find and count their night nests to try to determine the size of the group (the assumption being that each weaned gorilla makes only one nest per night), but these counts would often give vastly different results. After a heavy rain, many nests can be easily counted, since gorillas bend vegetation to separate themselves from the wet forest floor and often move in search of better spots throughout the night. Fewer nests can be counted in dry months, when the gorillas often sleep on the bare ground, leaving nothing but their dung and a few hairs behind. Consequently, nest counts could vary from as few as three to as many as eleven, leaving us wondering just how many gorillas were in "our" group. Within a few months of piecing together our observations we estimated that there were from six to eight individuals: the silverback, at least three adult females, one juvenile, and two infants. But then suddenly there was an abrupt change that made us appreciate the fragility of social groups.

• • •

November 15, 1999

Today the team with the gorillas called camp by radio to report some alarming news: Mlima is severely wounded, and most of his group is missing. From the description of the injuries, it sounds as though they may possibly have been caused by hunters, another gorilla, or a leopard.

My heart sinks, and I can't stop thinking about what might have happened and whether we could have prevented it. The previous evening, we had left the group as usual, just before nightfall, as the gorillas were making their night nests. We had sensed nothing unusual. Did Mlima engage in combat with another silverback? Did he lose his females? Could a leopard really have attacked an almost 200-kilogram silverback? Or did Mlima try to defend a smaller individual of his group being targeted by the leopard? Are there hunters in the park??

• • •

The next morning, while a team inspected the area where Mlima was wounded, I went to check on the extent of his injuries. I found Mlima huddled in a ball, moaning in pain. Standing several meters away, we immediately saw a large open wound above his left eye extending across his forehead. Then we noticed three long and narrow scars behind his ear and another long one behind his shoulder.

Evidence at the nest site revealed that a fierce battle had occurred: trails of blood crisscrossed the trampled vegetation, impregnated with the musky smell of silverback. On a vine the team noticed fresh scratch marks: evidence that a leopard had just been there. At least we now knew there had been no human hunters in the area. But was a leopard really to blame, or had there been a fight between two silverbacks, with the leopard arriving later?

We'll never know exactly what happened that day. Part of the frustration and the beauty of working in the field is that we cannot control, monitor, or foresee the events that nature has in store for us. Over the years, this will certainly not be the sole unsolved mystery in the forest, but will certainly remain one of the most puzzling for us.

My only relief came from excluding the possibility that Mlima had been wounded by human hunters. In the previous months, we had put a lot of effort into discouraging people from using the area where gorilla habituation was taking place. We had worked in close contact with the anti-poaching conservation teams, while education campaigns were being carried out to explain conservation and the long-term use of forest resources, including gorilla tourism.

Everybody knows that approaching wounded animals can be dangerous: they might react more aggressively because they cannot run away. When we approached Mlima, he was so weak he could barely move. There he was, sitting in pain, facing us. Oddly, we felt as if this time *he* was studying us. It is my belief that during that time he realized that, had we been a danger to him, we could have easily harmed him in his weakened state. Once recovered, Mlima would still occasionally charge us, but he seemed to have lost his fear of humans. In a way, his misfortune worked in favor of accelerating the habituation process, a bittersweet way of making a big step forward.

Amazingly, Mlima recovered quickly from the injuries but his group was marked permanently by the departure of most of its members. Only three remained: Mlima, Matata, an adult female, and Ndimbelimbe, their juvenile son. Was somebody else injured? Had they all left with another silverback? Another set of questions that would be left unanswered.

For Ndimbelimbe, now the only young gorilla of the group, our team was definitely a welcome distraction. Eventually, when his parents did not reprimand him, we would have to discourage his attempts to play too close to us. Being very close to wild gorillas may be tempting, but it bears the heavy risk of disease transmission. Wild gorillas have not been exposed to many of our diseases (such as the common cold), and they thus do not have the antibodies to fight them. The price of a thrilling experience or a close-up photo could be the life of an endangered gorilla.

While gradually gaining the trust of this small family, we started to organize the tourism program. Logistics in a remote rain forest are more than just details; we had to make sure the tourists arrived at the right time and the right spot to see the group each day. Above all, to guarantee the well-being of all involved, gorillas and humans alike, we had to set clear guidelines and make sure they were followed. The rules for visiting gorillas at Dzanga-Sangha include: keep a safety distance of ten meters between humans and gorillas; no visits while showing symptoms of sickness; no littering; covering the mouth while coughing; and using the toilet in camp before going into the forest. Unsurprisingly, demand by tourists to observe gorillas in the wild is high. To see western lowland gorillas in their forest environment may be a tough adventure, requiring long walks through thorny vegetation in an exceptionally humid climate, but it is an overwhelmingly rewarding experience finally to be able to see gorillas up close.

• • •

December 1, 2001

I am sitting on a log, watching Mlima feed on termites. A crunching noise comes from behind us. A flashback of yesterday's close encounter immediately comes to my mind; I jump up and scream: "Elephant!" The trackers are laughing, and I realize with some embarrassment that the noise was not made by an elephant but by an approaching group of tourists! To my surprise, Mlima barely lifts his head for a glimpse at the commotion and then resumes breaking open the termite mound. Finally, encountering humans on a daily basis has become somewhat mundane for Mlima—great news for us. While the visitors juggle their binoculars, cameras, and video recorders, I look at the proud smiles on the trackers' faces, and I too smile. Nyele, in his Ba'Aka-style basic French, tells the tourists of the many years he has worked to habituate the gorillas and Ngombo recalls his first glimpse of Mlima, when he still doubted gorillas could be habituated at all. We decide to celebrate and invite the tourists over for a special pizza party, combining Italian cuisine with the Ba'Aka's wild drumming and singing. Quite a mix of cultural experiences!

• • •

The Ba'Aka have an amazing and intimate knowledge of the forest—and much more. Their skills and ability to interpret the subtle signs left behind by animals in the forest are what made it possible to follow and habituate gorillas in the dense forest of Dzanga-Sangha. Their culture, humor, and joie de vivre have made for many eventful and unforgettable days during all the years I have spent in Central Africa. In a way, I also like to think of what the Ba'Aka have learned from our program. At the start no tracker believed in the habituation process. The thought of one day being able to sit next to a silverback and his group while they fed or rested, ignoring our presence—well, it was just inconceivable. Yet, with time, not only did it come about, but people who used to see gorillas either as prey or danger now referred to each one of them as an individual, with his or her character and a story to be told.

We first opened the Gorilla Tracking Program to tourism in October 2001. Before starting our program, Dzanga-Sangha was famous for the most exceptional forest clearing in the Congo Basin: Dzanga Bai. *Bai* is the local term for a natural clearing, whose salt-rich soil attracts the secretive forest elephants and other wildlife, including bongos (the biggest forest antelope), buffaloes, and sitatungas (marshbuck). Gradually, the gorillas are becoming Dzanga-Sangha's main attraction, a big reward for those who worked many long hours in the forest "chasing shadows." In less than three years, over 500 people from all over the world had come to this remote region in Africa to meet the Munye group. Mlima and his small family have played a leading role in four documentaries and have appeared in several others. Volunteers and researchers have joined our program to help the tourism program and carry out research projects.

• • •

July 1, 2004

It is the end of the day, and Mlima and Ndimbelimbe prepare for the night. Since Matata left for another group the year before, father and son have been ranging over a small area, spending their days quietly and often nesting down early. We cannot help but wonder what the future of this odd group will be. How long will Ndimbelimbe stay with his old, half-blind father? Suddenly a dark shadow approaches, and Mlima bursts in with a loud chest-beat. I cannot believe my eyes: it is a female! Who is she and where does she come from? Another forest mystery . . .

• • •

When Matata left the group, leaving behind her own son Ndimbelimbe, we were not surprised. Mlima was getting old and upon encountering other males, he would often silently drag his little group away to avoid any confrontation. He was certainly not the best candidate to father her next infant, and Ndimbelimbe was old enough to fend for himself. Matata joined another group after observing its silverback threatening and challenging Mlima. Ndimbelimbe stayed with his father: he could not have followed Matata into the new group, because the silverback would not tolerate the arrival of another male, albeit a young one. What we were not ready for was Mlima's reaction to her departure. For ten consecutive days, Mlima was inconsolable. He would spend his days hooting in a way he had never done before; it sounded more like a heartbreaking plea than the powerful vocalization of a threatening silverback. With the passing days, the frequency of his hooting decreased, and gradually he settled into a very quiet existence, almost as if he had lost interest in life.

The arrival of the new female, Samba, was a moment of intense happiness that we were not expecting to witness. We do not know which group she came from and why she had chosen to join old Mlima. From the day of her arrival, the change in Mlima was drastic: the slow, sleepy, half-blind old silverback we knew so well suddenly turned into a boisterous, loud energetic male with a new purpose in life: to keep this female by his side.

But things did not work out for Mlima and Samba: a month after her arrival, another male set his sights on her. After a series of interactions, the two males engaged in a final combat, in which Mlima lost Samba and all his strength. On the morning of August 1, 2004, we found Mlima near his night nest severely wounded on his head, chest, arms, and back. Ndimbelimbe was nowhere to be found, and we feared the worst. Months later, we met him alone in the forest: he had lost the use of one arm and we could only speculate that he had been injured trying to protect his father and keep Samba with them.

It was immediately clear from the extent of his injuries that this time Mlima would not be able to survive. As I watched him through my tears, lying on the ground, breathing heavily, I felt an immense sadness. At the same time I felt so grateful to this beautiful gorilla who had allowed us to witness the joys and despairs of his life.

The whole team was devastated by Mlima's death; he had played such a central part in our lives over the past six years, as we followed his group and their adventures on a daily basis. Somewhat to my surprise, on my Sunday trips to the villages, I received countless messages of condolence. We had not been the only ones following the saga of Mlima and his family. As a part of our program, the local villagers received free visits to the Munye group once a week, a successful approach toward a wider participation in and understanding of the tourism program. At night, when people gathered around the fire, the trackers would often act out the adventures of the magnificent, gentle silverback named Mlima for a captivated audience.

As we worked at habituating other groups, inside us a new strength was growing. The opportunity to know Mlima and his small family had been a tremendous inspiration for all those who met him and given us the motivation to carry on our work. The tourism program continues successfully with a second group, called the Makumba group, while others are in the process of being habituated. The lessons we learned with Mlima were not in vain, and the Makumba group continues to teach us more each day.

FURTHER READING

Blom, A., C. Cipolletta, A.M.H. Brunsting, and H.H.T. Prins. 2004. Behavioral responses of gorillas to habituation in the Dzanga-Ndoki National Park, Central African Republic. *International Journal of Primatology* 25:179–96.

Cipolletta, C. 2003. Ranging patterns of a western gorilla group during habituation to humans in the Dzanga-Ndoki National Park, Central African Republic. *International Journal of Primatology* 24: 1207–26.

———. 2004. Consistency and variability in the ranging patterns of a western gorilla group *(Gorilla gorilla gorilla)* at Bai Hokou, Central African Republic. *American Journal of Primatology* 64:193–205.

Harcourt, A.H., and K.J. Stewart. 2007. *Gorilla society: Conflict, compromise, and cooperation between the sexes.* Chicago: University of Chicago Press.

Masi, S., C. Cipolletta, and M.M. Robbins. 2009. Western lowland gorillas (*Gorilla gorilla gorilla*) change their activity patterns in response to frugivory. *American Journal of Primatology* 71:91–100.

TEN

Among Silverbacks

MARTHA M. ROBBINS

July 2001

I was hot, sweaty, dirty, muddy, wet, and very out of breath. Following a heavy breakfast of beans and potatoes at camp, we'd been walking for two hours on a roller-coaster ride, up-up-up and down-down-down, up-up-up and down-down-down through the never-ending series of hills of Bwindi Impenetrable National Park, Uganda. Walking, or even simply remaining standing, in this forest is not easy because of the dense network of plants: roots, stems, thorns, and thickets make us trip or slip, or just basically trap us in a maze of vegetation. The field assistants ahead of me tried not to notice that I'd fallen down for the third time that morning and politely slowed their pace to mine.

Then we noticed that a clump of vegetation was shaking vigorously thirty meters ahead. Experience told me that a gorilla was feeding there, and that it was not two plants having a wrestling match. We slowly advanced toward the two-meter-high shaking plants and a glimpse of silver revealed that it was Zeus, the dominant silverback of the group. There was no King Kong drama to mark our arrival. Zeus simply threw us a glance and continued to crunch loudly on some tasty piece of vegetation. Immediately, all my agony of the typical morning hike faded away, even my feelings about the plants.

But I knew I couldn't settle into the routine of observing the morning feeding session. First, I needed to confirm who was in the group, because five days previously, Rukina, a young silverback, and Tindamanyere, a young teenager-like female, had been reported missing. At least we knew nothing bad had happened to them, because the previous day, field assistants had gone back to where they had last been seen and managed to follow their trails in the vegetation until they were found, approximately two kilometers away from the bulk of the group. Their departure from the group occurred exactly twenty-eight days since they had last mated, so I assumed that Tindamanyere was having regular estrous cycles, and that their disappearance was likely to be sex-related. After about thirty minutes of crashing through the thick vegetation, I determined that all fourteen members of the group were there except for the two young lovers. As I started to conduct my observations on Zeus and the other five adult females of the group, I wondered whether Rukina had decided it was better to split off from the group with only one female rather than continue the battle with Zeus for leadership.

Rukina had already tried the strategy of going solitary and then changed his mind. He had spent six months as a solitary male, but then he rejoined the group, something never before recorded in mountain gorillas. His foray away from the group with Tindamanyere was also something not previously observed.

I didn't have long to ponder scientific thoughts about the variation in reproductive strategies of gorillas, because I suddenly noticed that Tindamanyere was surreptitiously approaching Zeus, as though she had never been missing. As she sat down to feed, I glanced around the hillside from which she had come and saw a large patch of silver and black that had to be Rukina bounding downhill. Zeus noticed Rukina as well and strode to meet him. In a moment of extreme anthropomorphism, I debated over whether they looked more like swaggering cowboys or sumo wrestlers preparing for a showdown.

Rukina stopped about twenty meters uphill from Zeus and struck an

ARE TWO MALES ONE TOO MANY?

In all mammals, there is intense competition among males to reproduce with females (see also text boxes in chapters 4 and 8). This competition is responsible for a great deal of the diversity we see in primates, including not only differences in social behavior and grouping patterns but what are called "sexually dimorphic traits," meaning the ways in which males and females differ physically. Gorillas are one of the most dimorphic primate species, with the males being nearly twice as large as females. The males also develop large crests on their heads, and the saddles of their backs turn silver when they reach maturity. All males become silverbacks, but not all males become dominant leaders of groups. Presumably, the reason why the males become so large is that bigger males can out-compete smaller ones.

Whether groups of primates contain one or many adult males is dependent on many factors, including ecological conditions that influence the number of females per group, dispersal patterns of females, the costs and benefits of forming alliances with other males, and the ways in which males compete with each other. These differences in males' ability to monopolize reproductive opportunities lead to variation in the proportion of offspring sired by dominant versus subordinate males, which is what biologists refer to as the degree of "reproductive skew." Groups of many other primates, including chimpanzees and bonobos, contain several adult males, typically with a dominance hierarchy and an alpha male leader, but gorillas tend to be more competitive, leading to more extreme monopolization of leadership.

If you read any of the early literature about gorillas, groups are described as containing only one silverback. So I was very surprised when I first went to Rwanda in the early 1990s to find that all three of the groups studied at the Karisoke Research Center at that time contained more than one silverback. Hadn't these guys read the books? Was something going wrong in these groups?

While the majority of gorilla groups contain only one silverback, some contain two. Males have two strategies to choose from in their ▸

▸ attempts to become a dominant male. As male gorillas reach maturity, many emigrate out of the group into which they were born, becoming solitary with the aim of attracting females from other groups to then form a new group. Alternatively, a mature male gorilla remains in the natal group, waiting for the older male, sometimes but not always his father or a half-brother, to concede the leadership of the group or to die. Initially, it was thought that multi-male groups were only a temporary situation, and that the young males emigrated shortly after they matured into silverbacks. However, long-term studies at both Karisoke Research Center in the Virunga Volcanoes region and in Bwindi Impenetrable National Park revealed that some groups have this multi-male structure for a large proportion of the time, with some silverbacks living in the same group for more than a decade. This strategy of remaining in the natal group is also referred to as "queuing," or waiting in line to attain the alpha position. Being number two in a queue may not be much of a line for waiting, but on rare occasions in very large gorilla groups (over thirty gorillas) there are three, four, or more males. Genetic studies have shown that as many as 15 percent of a group's offspring may be sired by subordinate males, which is clearly better than what they would achieve if they were not in a group at all. This also shows that dominant males do incur some costs of having other males in the group, leading to several questions: Do males voluntarily emigrate or are they kicked out? If males don't voluntarily go, is it too costly for dominant males to kick these other males out? Does the dominant male gain some benefits if he permits younger males to stay?

Additionally, is a male likely to be more successful by emigrating and attempting to form his own group or by queuing? The answer to this question depends on where you are. For mountain gorillas, it appears that queuing is often the better approach, and as many as 40 percent of groups may be multi-male. In contrast, emigrating is believed to be the best approach for western gorillas, since nearly all males do it and multi-male groups are extremely rare. Why the difference? We still don't have all the answers, but most likely the number of females in the group influences a male's decision, and the number of females is ▸

▸ likely influenced by the availability of food. Mountain gorillas live in habitat with extremely abundant vegetation, so they don't need to invest much energy in traveling to meet their nutritional needs, whereas western gorillas live in an environment with less easily accessible vegetation and need to put more effort into getting enough to eat. These ecological differences may enable mountain gorilla groups to become larger than what is optimal for western gorillas, thus favoring the formation of multi-male groups. Males are also likely to emigrate out of groups where they have only close relatives, so as to avoid inbreeding.

This variability in the number of adult males per group among different populations of gorillas has implications for the types of social relationships in the groups, as well as for the stability and longevity of groups. If the only male in a group dies, the group disintegrates, because the females disperse to other silverbacks. If the dominant male in a multi-male group dies, a younger, subordinate male may be able to take over leadership immediately and retain the group. If a group doesn't disintegrate, there is no risk of infanticide, which is a benefit for females in multi-male groups. This may be one way in which females can influence the occurrence of multi-male groups, but we haven't yet seen evidence that females prefer to join multi-male groups. Furthermore, given this advantage, it is puzzling that western gorillas almost never form multi-male groups. Ultimately, it likely comes down to competition between males.

Being a silverback comes naturally to all males that reach maturity, but being a successful silverback is extremely difficult. Half of all gorillas born are males and half are females, yet groups have an unequal adult sex ratio since they contain only one or sometimes two adult males but typically three to five or more adult females. What happens to the rest of the adult males? Of the males that emigrate, many never succeed in forming their own groups and remain solitary for years. Given the social nature of gorillas, perhaps an equally appropriate name for these "solitary males" or "lone silverbacks" is what many Ugandans call them, "lonely silverbacks."

impressive quadrupedal pose. He stood in a strutting position, with his legs and arms stiff and extended, his head tilted slightly upward, and his lips pursed. He stood sideways from Zeus to show the full length of his body and the size of his head. He threw furtive glances at Zeus, never making direct eye-contact. Soon Zeus was standing in the exact same strut-display position, parallel to Rukina, watching his opponent out of the corner of his eye.

As they sized each other up, I tried to figure out where the best place was for me to watch. The gorillas were in a valley, surrounded by hills on all sides, creating the impression that we were in the bottom of a bowl. I hoped, given that the only alternative was to move uphill, that they would remain in the valley, where there was some semblance of flatness and it would be somewhat easy to maneuver around them. No such luck. We were going to the rim of the bowl, a few hundred meters away. Rukina suddenly ran perpendicular to Zeus, so as to not lose any of the advantage of being uphill, and at the same time not actually to approach him any closer. In response, Zeus immediately charged uphill to be slightly above Rukina, but also at an angle to him so as to maintain a ten-meter distance between them. As the silverbacks ran, they retained their deliberate, stiff-legged posture. When they stopped, they resumed standing in their strut-display position at an angle to each other so as to best show their silver backs, muscular bodies, and large head crests. The males stood parallel to each other for several minutes until Zeus charged further uphill at an angle to Rukina, only this time including a loud chest-beat in his display. Rukina quickly followed suit, while continuing to retain the ten-meter invisible boundary between them. Soon the valley was echoing with their chest-beats.

I wanted to follow my rule of staying uphill rather than downhill from charging silverbacks, but it was impossible, because the males continued to run back and forth at each other, weaving their way further uphill at a faster pace than I could manage through the tangle of plants. How could anything that weighed over 300 pounds actually run uphill in such a graceful manner through such thick vegetation, I wondered, as

I disentangled my foot from a snarl of vines and climbed up part of the slope that was much steeper than the average staircase. There was no way that I could get ahead of them. My two field assistants were warily following behind me. Given a choice, they would avoid any neighborhood in which silverbacks were fighting, but they thought that part of their job was to "protect" me. They hesitantly pleaded with me to be careful, reminding me that silverbacks can be dangerous. I courteously agreed, but continued to slowly inch uphill after Rukina and Zeus. I knew that the last thing those two silverbacks had on their mind was me, and that I basically just needed to stay out of the cross-fire. As I struggled uphill, I could see that the males were taking a pause and the rest of the group was slowly meandering uphill after them.

For the next thirty minutes, the show continued. We all were nearing the top of the hill and several of the adult females were gathering around Zeus. Rukina was already on the ridge and I was lagging twenty meters below Zeus and the others. The next time Rukina ran toward Zeus, he broke through the invisible ten-meter barrier, but only to have Siatu and Kakumu, adult females who were loyal supporters of Zeus, immediately scream and run at him. Zeus also started to run directly at him, so Rukina quickly turned around, but instead of continuing uphill, he immediately ran straight toward me. I barely had time to move one step to the side when Rukina galloped past me, only inches away, with Zeus in close pursuit. Rukina stopped just a few meters below me. Zeus stopped shortly in front of me. My heart stopped. While I didn't want to be in this situation, I couldn't move anywhere anyway, because there was a solid wall of vegetation on one side of me and a wall of silverbacks on the other. If anything, my position exactly between the two males only epitomized my feelings of not choosing a favorite of the two; I certainly did not feel that I was qualified to be a referee. While I stayed planted where I was, Rukina turned around to quickly look at Zeus and sat down, somewhat deflated. Zeus glared at me, then at Rukina, then he turned around and purposefully walked back up to the waiting females, having finished his business with the amateurs. No early retirement for him.

The drama was over for the day due to the important task of feeding, which occupies about half of the daylight hours for gorillas. Rukina moved away to eat, occasionally looking uphill to see where everyone was. Zeus also began to feed, with Siatu, Kakumu, and various juveniles and infants nearby. However, the feeding soon came to a stop when what started as a few heavy rain drops intensified into a downpour. Rukina had no trouble finding a dry spot under a large tree—alone. Further uphill, Zeus also found a tree to sit under, but there wasn't space underneath for everyone. Tindamanyere hesitantly walked toward Zeus's tree but was quickly rebuffed by Siatu, the dominant female, who aggressively grunted at her from Zeus's side. Tindamanyere slowly retreated to a few meters away from Zeus, resigning herself to sitting it out without any cover from the rain. Likewise, I resigned myself to sitting in the rain if I wanted to be able to continue to see as many of the gorillas as possible. I knew they would remain huddled in the exact same positions until the rain stopped, but watching gorillas do nothing in the rain is still better than watching nothing but the rain.

During the next two hours shivering in rain, I thought over the previous activities. As was the case in many of these skirmishes, the two silverbacks never actually made physical contact. However, the ten-centimeter bite wound on Rukina's back, which had not yet healed after two months, reminded me of the damage they were capable of inflicting. My mind was filled with questions: Would Rukina give up and emigrate again, in hopes of attracting females from other groups? Would Zeus slowly give up his dominance, but remain in the group, as I had observed years before with other mountain gorillas in Rwanda? Why did Tindamanyere appear to prefer Rukina, yet still mate with both males? When she had a baby, who would be the father? Was the support the other adult females gave Zeus helping to prevent Rukina from gaining the upper hand? When would this chapter of the soap opera end? When would it stop raining?

• • •

March 2005

Four years later, we reached the gorillas exhausted as usual, but quickly recovered as the gorillas revealed themselves, and I began to wonder what events the day would bring. It was sunny and warm, so I expected the gorillas to be taking cover underneath trees and shrubs during their midday rest session. Most of the group were munching away, sounding like a crowd of people eating celery. I spotted Rukina lying down in a small clearing under a tree on the edge of the group. He sat up briefly to see that it was us and then plopped himself back down on his stomach as if to make himself more comfortable on a well-worn sofa for an extra few minutes of morning snoozing. Relaxed new leader, I thought, while noticing that two factions of the group appeared to be moving away from him in opposite directions. Kabandiize, the two-year-old son of Tindamanyere and Rukina (confirmed to be his father by genetic testing using feces), similarly wanted to relax, but he knew that Rukina's back was the most comfortable seat in the forest. The infant pulled himself up onto Rukina's back and reclined, trying not to slip off. Rukina sat still, because he tolerated his son using his large body as a playground. Next, Matu, an elderly female, took a pause from feeding and purposefully sat directly in front of Rukina, subtly demanding to be groomed. Rukina sat up and complied. Kabandiize tumbled to the ground, but soon clambered onto Rukina again. Matu had previously solicited grooming from Zeus on a regular basis, and never tolerated Rukina within a ten-meter radius of her, but she rapidly changed her alliances following the final showdown between the two silverbacks.

The coup d'état had come in July 2004, following three years of on-and-off battles between the two males. The shift in power became only too apparent as Zeus showed obvious signs of aging and an increasing number of wounds, while Rukina continued to grow into his prime. The stakes were especially high when the final battle occurred, because three females were sexually receptive. The last weeks were particularly chaotic. At times the females didn't know which male to follow, and the group became somewhat scattered as the two males tried to move it in different directions. Often it was apparent that Zeus was trying to avoid Rukina by making sharp turns in direction and traveling when the group would normally have been resting. While we weren't seeing many fights, because they were rare, it was clear that Rukina was winning, since Zeus had such bad bite wounds on his head that his spectacular crest was no longer present—the top of his head was as flat as a female's. Several of the adult females also had bad bite wounds on their heads, suggesting that Rukina was trying to coerce them onto his side by showing his dominance and power.

On the last day that Zeus was in the group, he sustained an injury so severe to his leg that he could barely walk. The next day the group moved away with Rukina, leaving Zeus behind. Some of the juveniles, Zeus's sons, came back to sit with him for some time before finally following their mothers and Rukina. Zeus lived in solitude for five more months until he died as a combined result of his injuries and old age. Watching him in those final months was difficult. I had the greatest respect for Zeus as a leader, and I much appreciated that he was so tolerant of us entering into his world. I couldn't help but think that he had to be lonely and depressed in those months of solitude.

The death of any mountain gorilla is cause for concern. The entire population of Bwindi mountain gorillas, numbering only 300 individuals, could (if so inclined) comfortably fit into a Boeing 747 jet. Only one more jumbo jet would be needed to accommodate the other half of the entire mountain gorilla subspecies, the 380 living in the nearby Virunga Volcanoes region. But the mountain gorillas can't fly anywhere. They remain in their small island of habitat, surrounded by nearly a million farmers, each living on only a few hundred dollars per year. The gorillas are threatened by habitat encroachment, poaching, and risk of disease transmission from humans, as well as living in a region of the world that has been marred by political instability and war. Yet mountain gorillas represent what is perhaps one of the few conservation success stories. Thanks to massive efforts by several conservation organizations, an ecotourism program that generates millions of dollars per year, and increased protection for their habitat by local park authorities, the two gorilla populations have increased in size in recent decades. They offer a glimmer of hope for ape conservation across Africa.

The demise of Zeus was only a natural part of the lives of the gorillas, but with such a small population, every individual is extraordinary. And being a silverback is not easy. Zeus was more successful than many males, some of whom never obtain leadership of a group and live for years as solitary individuals. We knew that he had been dominant for almost a decade and had fathered several offspring. Zeus was a special gorilla because he was able to live out his natural life despite the severe threats to the small population in which he lived. It would likely be years before we knew how successful Rukina would be, and we hoped that it would be possible to observe his full life.

While Rukina, Matu, and Kabandiize continued their impromptu early rest session, we heard screaming from not very far away. Rukina sat up and looked to see if his intervention was needed. Siatu and Tindamanyere were quarrelling over a feeding spot, despite the fact that they were surrounded by equally good places to eat. They continued to scream at each other, with their faces only inches apart, until Rukina gave a short, aggressive admonishing grunt. Tindamanyere conceded and walked toward Rukina, while Siatu began feeding victoriously. When Tindamanyere was halfway to Rukina, she was nearly run over by two ten-year-old males, Marembo and Sikio, who were playfully chasing each other. Rukina also gave them a brief grunt, but they simply hurled themselves at each other and tumbled out of sight downhill. Tindamanyere finally reached Rukina and began to groom Kabandiize. Their contentment with the day was contagious.

As I watched them all rest together and listened to the blackbacks play as if there weren't a care in the world, I couldn't help but wonder what the situation would be like in a decade. I was doubtful that Marembo and Sikio would both be tolerated in the group by Rukina when they became silverbacks in a few years' time. It was hard to imagine little Kabandiize, who was just barely larger than Rukina's head, as a silverback, but would he be fortunate enough to lead a group? Regardless of who won in the silverback competition, I felt that as long as conservationists and researchers never became as complacent as the gorillas appeared to be at this moment, there was some room for optimism about their future.

FURTHER READING

Harcourt, A.H, and K.J. Stewart. 2007. *Gorilla society: Conflict, compromise, and cooperation between the sexes.* Chicago: University of Chicago Press.

Nsubuga, A.M., M.M. Robbins, C. Boesch, and L. Vigilant. 2008. Patterns of paternity and group fission in wild multimale mountain gorilla groups. *American Journal of Physical Anthropology* 135:265–74.

Robbins, A.M., and M.M. Robbins. 2005. Fitness consequences of dispersal decisions for male mountain gorillas *(Gorilla beringei beringei). Behavioural Ecology and Sociobiology* 58:295–309.

Robbins, M.M. 2008. Feeding competition and female social relationships in mountain gorillas of Bwindi Impenetrable National Park, Uganda. *International Journal of Primatology* 29:999–1018.

———. 2009. Male aggression towards females in mountain gorillas: Courtship or coercion? In *Sexual coercion in primates: An evolutionary perspective on male aggression against females,* ed. M.N. Muller and R. Wrangham, 112–27. Cambridge, MA: Harvard University Press.

———. 2010. Gorillas: Diversity in ecology and behavior. In *Primates in Perspective,* ed. C.J. Campbell, A. Fuentes, K.C. MacKinnon, S. Bearder, and R. Stumpf, 305–21. 2nd ed. Oxford: Oxford University Press.

ELEVEN

The Diversity of the Apes

What Is the Future?

CHRISTOPHE BOESCH

When I first drove from the capital of Côte d'Ivoire, Abidjan, to my field site in the Taï National Park in 1976, the road turned from paved tarmac to dirt some 300 kilometers from my goal, and the forest was visible from the road in many places. I still remember with emotion having to stop to let a mother chimpanzee and her two infants cross the road when I was still about 100 kilometers from the park. Nowadays, when I am driving to my field site, the dirt road starts only 80 kilometers before reaching the park boundary and I cannot see a single piece of intact forest until I reach the signs indicating the park entrance. Over the past twenty years, the country has been able to keep its position as the number one producer of cocoa in the world and the third of coffee, but this came with a price. The rate of deforestation has been one of the highest on the continent, and the resulting decrease in rainfall in the whole region has forced millions of people to migrate toward the south of the country, because it has become too dry to grow cash crops in the north. The fight for fertile lands has resulted in the longest civil unrest in the history of the country. As a consequence, the chimpanzee haven that Taï National Park previously was, with all its unique West African animals, has become an isolated island in the middle of a huge cocoa, coffee, and oil palm tree plantation. However, we recently estimated that in many other parts of the country that are not as well protected as Taï National Park, the chimpanzees and the forest have suffered to the degree that some 90 percent of what was present twenty years ago has disappeared.

When I visited the mountain gorillas in the Bwindi Impenetrable National Park, Uganda, the obvious signs of incredibly high human population densities completely surrounding the small park and the use of all unoccupied pieces of land to grow potatoes, beans, and cabbage gave a dire perspective on what could happen in the years to come in other regions of Africa where great apes live still in relative safety. In the past decades, even the chimpanzees and gorillas living in northern Congo, one of the least populated regions of Central Africa, have seen advancing deforestation and the incredibly rapid development of human settlements and infrastructure accompanying logging operations in the previously untouched regions adjacent to Nouabalé-Ndoki National Park. Such development attracts many hundreds of workers and their families to remote logging villages and results in increased pressure on the forest in terms of both hunting bushmeat and extracting firewood.

I am not arguing that economic development is bad; my point is that it needs to be planned so that it can progress in tandem with conservation of nature. In Côte d'Ivoire, one of the most pressing issues in the past two decades has become water. All elderly people in the villages will tell you that rainfall was much more abundant and predictable in the past than it has become nowadays. One of the direct consequences of this is that cocoa, coffee, and palm oil yields have dramatically decreased. Many villages in the center of Côte d'Ivoire, which used to be the heart of the cocoa production region in the 1960s, have been totally abandoned and people have moved toward Taï National Park in the west of the country in an attempt to follow the water and be able to continue to grow cash crops. In a vicious circle, more people arrive in the forest regions because they are unable to survive with the bad yields elsewhere; this increases the pressure on the forest, and if it is cut, yields will decrease again. However, the forest is not endless. The more forest you cut, the more rainfall decreases and the more desperate the situation becomes for farmers. Sustainable economic development will only work if the long-term interest of nature is included in the planning.

Sadly, this is not always understood, and too many people still believe that nature conservation is detrimental to human development. In reality, the two are complementary more often than one would think. All the contributors to this book hope that conveying their fascination with the great apes to the public will help make it clear that by saving apes we are also saving humans and helping our own future. The gorillas of Dzanga-Sangha (chapter 9), and the ones of Mbeli Bai (chapter 8) are already part of the ecotourism sector, which is one of the most rapidly developing branches of the tourism industry worldwide. Some of us are working actively in such projects because we feel this is one way by which the local human populations will directly profit from the presence of great apes, and that this development could be one of the strongest solutions in the face of an ever-growing human population in Africa. Studies with the mountain gorillas of the Virunga Volcanoes region in Rwanda have shown that this population, highly threatened by a very dense human population surrounding it, has been faring well despite dramatic political instability over the past two decades, and that ecotourism and monitoring of the gorilla population were instrumental in making that possible.

The intimate glimpses in this book into the lives of some eight different groups of apes in Africa have been made possible by years of painstaking effort on the part of dedicated field-workers. These groups, however, represent only a tiny fraction of all the apes that are still alive in the remote forests of the continent. We can only wonder about how many more aspects of their lives we might discover if we could observe them. We hope that by having brought you on a journey into the world of the African great apes, we shall be able to enlist your interest in trying to protect them.

However, effectively protecting the apes will only be possible if we learn more about them, because we still are missing key information. For example, we still don't know how many of them live in Africa, where exactly they live, and what specific threats they encounter.

NUMBERS ARE MORE THAN JUST NUMBERS: ARE THEY GOING TO SURVIVE?

In 2003, Martha Robbins and I explored the forests of central Gabon (see chapter 1), looking for apes in the newly created Ivindo National Park. The biologist Mike Fay had crossed this forest a few years before during his "mega-transect" walk across central Africa, and told us that he found a large number of gorillas in Ivindo and that they were the least scared of humans of any apes in the region. These were great conditions for establishing a new field site. Unfortunately, as we walked for hours, we first realized that there had been extensive logging in the northern part of the park and all the largest okoume trees had been cut and removed since Fay had been there only a few years before. For the weeks we walked through that forest, we met a few groups of chimpanzees, but found only one single fresh trace of a gorilla. Where had the gorillas gone? How is it possible that within the few years since Fay's passage, the gorillas had all disappeared? It seems unlikely that this could be due to logging. We still have no clear idea of what happened, but we now think that they might have fallen victim to the Ebola virus.

Eight years ago I learned a shocking piece of news. I asked a colleague if we could develop a model to predict the viability of gorilla populations across Africa over the next ten years, only to be told that we had no precise idea of how many apes currently survive in the wild. Journalists also have a very hard time believing this and always ask the same question: How many gorillas and chimpanzees live in Africa? How is it possible that after decades of investing in the conservation of great apes, we still don't know how many of them there are, where they live, and by how much they may be declining? Sometimes we don't even know what is contributing to their decline. Imagine an automobile manufacturer that did not know how much it cost to build a car, how many cars it produced in a year, or how many cars it sold. Impossible, you would say. But something like this is the current situation in the world of great ape conservation.

Several factors contribute to this situation. First, donor agencies prefer to give money to conservation projects that might "save apes," rather than give money to estimate how many there are and where they are. Investing money in a project that simply provides a number is not considered to be particularly exciting. In the long run, this has proved to be a trap, because having baseline information on population sizes is crucial for monitoring both natural changes and those that result from conservation activities. Second, apes live deep in the most remote parts of the African jungle. Access is often terrible, visibility is very low, and the vast majority of wild apes are unhabituated to humans. Therefore we cannot count them directly and have to rely on the traces they leave behind, such as the nests in which they sleep or feces on the ground. Unfortunately, these indirect traces can be used reliably only if you know how many are produced by an ape each day and how quickly such traces disappear. This has proved remarkably difficult to figure out because, given variations in the trees in which apes build nests and the amount of rainfall, these variables differ depending on where you are and when you are there. The more it rains, the quicker feces and nests tend to degrade, while the content of the feces and the tree used to build a nest also play a role. Some people thought we could neglect these details and base our estimates on the general pattern of vegetation found in a forest, but, sadly, poaching and diseases like the Ebola virus can simply wipe out ape populations in a forest. There may be cases of the famous "empty forest syndrome" where a perfect forest exists but is empty of animals. Therefore, we need to get into the forest and look for great ape signs to make an estimate and not rely only on satellite imagery of forest cover and vegetation patterns.

Now we are in the midst of a race to change this situation and accurately count apes at the same time as we try to protect them, which is far from ideal. But at least in past decades we have learned much about the ecological conditions that are important for apes, and we therefore have a better understanding of where to expect to find them. We can also profit from the scientific progress that has been made in recent years to better estimate the number of apes in an area using the indirect methods described above. Moreover, the great progress made in remote sensing technology, including the extensive use of satellite pictures and GPS technology, will make this surveying effort of wild ape populations much more accurate. And finally, new survey techniques using other modern technology such as motion-sensor video cameras, audio recording devices, and genetic methods are being tested. Thanks to these advances, the mystery about the real number of great apes in Africa may be cleared up in the not too distant future.

Even with the problems of knowing how many apes are in a particular region, we have been able to document some dramatic declines. A study conducted in 2008 concerning the remaining populations of western gorillas in Central Africa came to an appalling conclusion: From the scant data available it was obvious that the number of gorillas in Gabon, Congo, and Cameroon was declining so rapidly that it was decided to upgrade their conservation status from "endangered" to "critically endangered." They not only suffered from deforestation and poaching but, in addition, the Ebola virus hit them very hard, decimating some of the largest and best-known strongholds of the population in such national parks as Odzala in Congo and Minkebe and Ivindo in Gabon. Since we know that about 80 percent of the remaining apes in Central Africa live outside protected areas, it is more urgent than ever to help apes that live in mining and forest concessions.

For the past fifteen years, war has been raging in the eastern Democratic Republic of the Congo (DRC), including many areas where gorillas live. Gorillas have been among the innocent victims. Sadly for the gorillas, coltan (columbite-tantalite) ore is found in high concentration in their habitat, and tantalum extracted from it is in high demand for our cell phones, computers, DVD players, and video game systems. The export of coltan has helped to finance the terribly deadly civil war rampant in that region. Under such pressure, the gorillas of the Kahuzi-Biega National Park and most likely other regions of the eastern DRC have suffered a dramatic decline. Although this is not inevitable, because mining can be done in sustainable ways, and some mining companies are ready to invest money to do so, it is virtually a fantasy for that to happen in a region torn by civil war. The growing realization that the resources of our planet are not endless, and that we need to use them in a sustainable way if we want to guarantee a future for humankind, has led to many advances in environmentally friendly extraction techniques not thought possible only a few years before. But we are faced with a race, and we need to be serious about it if we want our cousins in the forest to have a chance.

THE POSSIBLE FUTURE OF GREAT APE RESEARCH

While the contributors to this book are clearly interested in the fate of the apes, we are also deeply curious to learn what makes an ape an ape, and we realize that after years of following them, we are still far from really understanding all facets of what they are. Prior to forty years ago, almost nothing was known about wild apes, so there has been a huge leap in knowledge in recent decades. However, mysteries endure. We are trying as hard as we can to decipher various unknowns, but that does not seem to be enough. If I am to be honest with you, I have to admit that it is those mysteries that keep us going year after year in the forest, as though we were Sherlock Holmes in the jungle trying to move forward our understanding of the African forest darkness. Also, as is often the case in science, one discovery leads to even more questions.

Our quest is not only to understand the apes, but equally to learn more about ourselves. The apes are our closest living relatives, and as such they provide a reflection on our roots and about our shared history. Because chimpanzees use tools, we know tool use per se is not a uniquely human ability, but we know as well that chimpanzees do not build airplanes or cars, for example, and therefore it is important to be more precise about what the differences are. Because gorillas and chimpanzees invest years in raising their offspring, we know that prolonged investment in one's children is not uniquely human. Similarly, the fact that bonobos and chimpanzees hunt for meat means that hunting is not a uniquely human ability in primates. In other words, knowing more about the great apes, our cousins in the forest, helps us better understand humans. We need to study more great ape populations to get a more precise knowledge of their diversity in order to answer the question, "What makes us human?"

The main discovery we have made in the past twenty years about great apes is that they have an impressive ability to adapt flexibly to varying ecological conditions. This forces us to expand our research on wild populations into new locations if we want to truly understand them. Yes, I have studied Taï chimpanzees for the past thirty years, but that does not permit me to say how chimpanzees in Gabon or Congo or Tanzania behave. Just as people in Greenland or the Congo rain forest do not behave, speak, count, or dress like people in Paris, chimpanzees in Taï do not behave, hunt, use tools, or eat like chimpanzees in Goualougo or Gombe. Imagine how inaccurate it would be if we described all human behavior from around the world based only on what people from Fiji do.

A CHIMPANZEE IS NOT JUST LIKE ANY OTHER CHIMPANZEE

In May 1990, I was invited by Jane Goodall to study the chimpanzees of the Gombe National Park in Tanzania. I viewed this as a great honor, because like most chimpanzee observers, I was inspired by her pioneering work in Tanzania and her dedication in trying to understand our closest living relatives. I remember vividly the first day I encountered the chimpanzees of Gombe high up on the slopes of this lovely park, after having spent eleven years studying the chimpanzees in the Taï forest. I heard them calling above us, and with the Tanzanian guide we quickly moved ahead. After about twenty minutes we spotted Freud. To me, based on my work with Taï chimpanzees, he looked like a healthy young adolescent with a long body who would probably develop into a big male. But then my guide told me that Freud was already twenty years old and one of the largest adult males of the group. I could not believe my eyes. How could the chimpanzees in Gombe be so much smaller? But this was not the last of my surprises, and my six-month stay in Gombe was one of my richest experiences in learning about differences among chimpanzee populations.

A few days later, I was following Frodo, Freud's youngest brother and grandson of the famous Flo whom Jane Goodall describes so vividly in her bestseller book *In the Shadow of Man.* Frodo was the largest male in the group and by far the best hunter. Since I was there to try to understand if it was true that Taï chimpanzees hunted differently from Gombe chimpanzees, I spent a lot of days following this beautiful male. As Frodo was reaching a nice patch of Miombe forest, he heard some of his preferred prey, red colobus monkeys, calling and he rushed toward them. He found a group of monkeys in small trees bordering the savanna, where they were quite exposed and vulnerable. It seemed a preferred hunting opportunity to my Taï-trained eyes.

Frodo seemed to think the same as he looked at them intently from underneath. Quickly, however, a few large male colobus monkeys descended toward him and threatened him with aggressive calls. Frodo was more than three times bigger than they were and could easily have overpowered them, but to my surprise, upon being confronted by the males, he hesitated and started to scream loudly with fear. He screamed for a long time, climbing only two meters up a tree before coming down again, away from the threatening males. Attracted by his screams, after some minutes more chimpanzees arrived. I thought that the hunt would start, but the newcomers simply settled down to watch Frodo and the monkeys. Frodo tried again to climb toward the prey, but this time two screaming male monkeys jumped on his back and tried to bite him. Frodo shook them off, rushed down the tree, and ran away, chased by three monkeys. I have seen hundreds of hunts in the Taï forest, but I had never observed an adult male chimpanzee running away from a red colobus monkey.

Taï chimpanzees can be intimidated by threatening monkeys, but a male chimpanzee would invariably attack and often succeed in capturing such large males. I would never have thought that chimpanzees could be so scared of monkeys. How could such an incredible difference arise? This reminded me of how, when searching for the chimpanzees alone in Gombe two days earlier, I came unexpectedly close to a group of red colobus monkeys, and without hesitation some large males ran toward me making threatening calls. That had never happened to me in my years in the Taï forest. The small trees within the woodlands of Gombe, only ten to fifteen meters high, sometimes make interactions between monkeys and humans (or chimpanzees) so close that monkeys

HUNTING BY CHIMPANZEES

All known wild populations of chimpanzees have been seen to hunt for meat. In some populations, hunting is done opportunistically when small mammals and birds are encountered nearby; in other populations, hunting is a very common team activity. Chimpanzees in Uganda, Tanzania, and Côte d'Ivoire have been seen to hunt for large red colobus monkeys on average once every three days. Hunting is done predominantly by adult group members, and females have also been seen to hunt and team up with males. To capture their main prey, arboreal African monkeys, up to ten males search for them together. Once they locate monkeys, the chimpanzees will rapidly chase them in the trees until they can corner and capture one. When successful, they will share the meat, preferably between male hunters and with sexually active females, so that up to twenty individuals may profit from such a kill. Taï chimpanzees, which live in a continuous dense rain forest, organize themselves as a team and have been seen to perform different, complementary roles so as to increase the likelihood of a catch. "L'union fait la force," as we say in French: "Unity makes strength." Meat sharing supports teamwork in Taï chimpanzees, because hunters receive ▸

try to mob intruders before running away. In the forty-to-fifty-meter trees of the Taï forest, such a situation never occurs. A small ecological difference can lead to an important difference in behavior, as Gombe chimpanzees were often in a situation where red colobus would attack them, even when they were not in the mood for hunting. Once I was following another male in a deep forest patch and suddenly I saw, without any warning, an adult male red colobus jump from the tree above directly onto the chimpanzee's back.

Frodo was not only very keen about hunting, but he was also very brave in his method. When he was in his prime, he was a very successful hunter, able to overcome his fear of the colobus males and zig-

▸ more meat than others, with the best hunters getting the most, while in other populations, such as in Tanzania and Uganda, meat is shared mainly with individuals who have strong social bonds.

The acquisition of group hunting skills in Taï chimpanzees is a long and slow process. Eight-to-ten-year-old male chimpanzees will attempt to capture prey without any consideration of whether others are acting in concert. These young males are never successful, except at being chased away by the adult male monkeys. Adult fifteen-year-old males will wait for one another but still have some difficulties in knowing where to place themselves so as to complement others in the team. Finally, thirty-year-old males are able to place themselves perfectly during a hunt in coordination with the others and in anticipation of where the prey is going to flee.

In the past, humans have been very successful hunters, and chimpanzees are the only other primates to share our propensity to hunt for meat. However, chimpanzees have not been seen to incorporate weapons systematically into their hunts and, as a consequence, have to restrict themselves to smaller prey, weighing at most twenty kilograms.

zag his way through a group of them to reach the females and snatch their babies away. For years Frodo initiated almost all hunts, and his group hunted very regularly, routinely capturing as many as five or six monkeys during a single hunt. I learned to admire his hunting style, although it was so dramatically different from that of the Taï chimpanzees. I was amazed to see that such a small ecological difference in the height and density of trees could lead to a succession of behavioral differences resulting in one chimpanzee population (Taï) hunting in teams for adult prey, whereas Gombe chimpanzees hunted mainly individually and for young prey.

The primary lesson we have learned from decades of fieldwork with chimpanzees is that we need to be very careful when we talk about "the chimpanzee." We should realize that each chimpanzee population faces different ecological conditions and reacts to them in a particular fashion, so that we should specify if we are referring to the Gombe, the Goualougo, the Mahale, or the Taï chimpanzees. Each chimpanzee population may hunt for meat, but in a different way for the same or different prey; each may live in a fission-fusion society, but differences may exist in average party size and in how social the females are with each other; each population uses tools, but there are differences in the types of tools used and the level of complexity achieved. I could continue the list, and a similar list exists for bonobos and gorillas, but I want now to illustrate this point with the case of tool use.

WHAT MORE DO WE HAVE TO LEARN? TOOL USE AS AN EXAMPLE

Taï forest, Côte d'Ivoire,
September 1983

Sweaty and panting, I was approaching a group of loudly calling chimpanzees on top of a rare steep slope. I tried to be very silent, knowing that while excited chimpanzees are easy to approach at such moments, it was still the beginning of the study, and they would vanish in the forest as soon as they spotted me. I heard regular thumping sounds coming from a tree. What was happening over there high in the trees? It did not sound as if they were cracking nuts with a hammer, which sounds more like carpenters. This sounded more like a drum. Through the leaves, I was lucky enough to see a female holding a large stick that she was using to dip into the big entrance of a beehive. She regularly licked the end of the stick, which was heavily covered with honey. I didn't realize how lucky I was at the time, because I would see this only three times in thirty years. However, for hours and hours during those thirty years, I observed chimpanzees using hammers to crack nuts, sticks to dip for ants, leaf sponges to drink water, twigs to extract grubs, or branches to stab leopards. After all that I thought I was an expert on tool use in chimpanzees. How naïve I was!

Loango Forest, Gabon,
July 2006

As I pulled myself through a swamp, I heard some thumping in a tree on the other side. Extracting myself from the mud, I realized the sound was very deep and had continued for the ten minutes I was struggling to get out of the swamp. As I approached what I hoped would be a glimpse of a chimpanzee in this totally unhabituated population, the sound stopped and I had to search very carefully through the leaves to discover a faint movement on my left. After I was able to find a place with a better view, but still remain unseen, I discovered a female high in the tree working with a stick in such a way as to enlarge the hole at the entrance to a beehive in the tree. After some minutes of this work, she moved from the bees' nest into the branches to break off a thinner stick. Bringing it back to the nest entrance, she inserted the entire seventy-centimeter-long stick in the nest and then removed it to lick the honey dripping from its end. She continued to dip for honey for fifteen minutes before she became aware of my presence. After she left, I went under the tree to discover a heavy thick stick on the ground that she most likely used to produce the deep thumping sound. There was one large stick to break the nest open, one medium stick to enlarge access to inside it and finally a thin stick to collect the honey from the inside.... Three different tools used in a sequential order to eat honey. I had never seen anything so complex in the tool use by the chimpanzees of Côte d'Ivoire. I had to explore the forest of Gabon to realize how much more complex and diverse tool techniques could be in chimpanzees.

Now is a very exciting time in chimpanzee research: After fifty years of work on eastern and western chimpanzees, we finally are getting the first observations of the central chimpanzees, which are by far the most numerous and widespread of this species. Detailed observations started in Congo and Gabon only in 2004 (see also chapter 6), and we are already discovering new and unique techniques of tool use that we thought were not part of the capacities of chimpanzees. All science has calibrated the standard of "What Chimpanzees Are" on the woodland-dwelling chimpanzees of Tanzania, where Jane Goodall did her pioneering work, but these are a minority that doesn't live in "mainstream" chimpanzee habitat and represents a very small part of what chimpanzees can be.

Now, with research finally expanding to include more chimpanzees living in the rain forest, which encompasses more than 80 percent of the chimpanzee populations in Africa, the findings may revolutionize our knowledge on many aspects of chimpanzee biology. Tool use promises to be one of these areas. Central African chimpanzees seem to use numerous and complex tools for extracting honey, a high-energy food, and have surprised us by using tools to reach invisible underground resources, such as termites and honey.

Yes, humans are the prime tool users. But what is it exactly that humans can do that other animals cannot? All new observations on chimpanzees using tools are potentially of key importance to understanding what is special in our own ability to use tools. Solving mysteries about tool use in chimpanzees will open a new window in our understanding of ourselves. I want now to move to another domain where we see that great apes can provide answers to mysteries of their lives that at the same time could provide indications about ourselves.

CULTURE IS AN ABILITY SHARED WITH APES

Such differences in tool use and other behaviors among apes have been an eye-opener for science, because this has led to discovering culture in animals. Culture has always been considered one of the most important characteristics of humans, by which we have developed our unique technological achievements and at the same time such wonders as Michelangelo's *David* sculpture and Mozart's Requiem Mass. Since we are not all Mozarts, more modestly, culture is our ability to learn how to speak, behave, and dress from our parents or our friends. This ability to learn our specific group traditions from others is the center of our cultural ability. It is here that the impressive differences in behavior seen in wild chimpanzees have shaken one of science's central tenets on humanity.

When Frodo encountered a nest of driver ants, he stopped to cut a one-meter-long stick from a nearby sapling. After inserting it into the entrance of the nest and letting the soldier ants climb on it for about twenty centimeters, he turned the ant-covered end toward his mouth and then swept the stick through his left hand to collect all the ants in one pile, finally putting them all in his mouth. I was fascinated by what I saw, because I knew by now that when a Taï chimpanzee eats the same species of ant, it uses a much shorter stick, and once the stick is covered with ants, it puts the ant-covered end directly in its mouth and pulls the ants away through its teeth. Nothing prevented Frodo from using a short stick, or the Taï chimpanzees from using a long one and both their hands. I tested these different options in both locations myself. But over the decades, all Gombe chimpanzees have been seen to use the Gombe way of eating ants, while all the Taï chimpanzees have been observed using the Taï way. Very reminiscent, indeed, of the difference that Europeans eat rice with forks, while Chinese use chopsticks.

The more we learn about differences among chimpanzee populations, and the more we learn about such differences that are not explained by any ecological or genetic variability, the stronger the evidence suggesting the presence of culture in chimpanzees. In 1999, all the field researchers working on chimpanzees joined forces and wrote a paper entitled "Cultures in Chimpanzees," which was published in the prestigious international scientific journal *Nature.* This opened the way to the idea that culture is shared with other animal species. Since then a similar paper has documented similar cultural differences in orangutans.

New studies on apes are documenting exciting new facets of our cousins in the forest. Each population that disappears will decrease forever our ability to understand fully what apes can do and, therefore, our ability to specify what is uniquely human in our own behavior and how much we actually share with apes.

WHAT IS THE FUTURE FOR GREAT APES IN A HUMAN-DOMINATED WORLD?

Our fascination with our closest relatives should not distract us from the fact that we are in a race against time to prevent them from going extinct. Is the situation hopeless? If it were, we would not have written this book. Knowing that much of our fascination with the great apes comes from their similarities to us and from the lessons they teach us about ourselves, we hope that this book can contribute to awareness that will make people concerned about this issue. In the previous pages Zeus, Rukina, Winona, Brutus, Volker, and the others have revealed to you that we are not just following great apes, but individuals with personalities, and that those individuals bring us into the politics and societies of their everyday lives. Each of them has revealed something about our past. With time, they will help us to decipher the remaining mysteries of the forest and the mysteries of human origins.

But will we be able to decipher them? Will the rapid deforestation in Africa give us enough time? Will the conservation activities designed to try to save the African apes be able to reverse decreasing ape populations? Many of us who have seen the effects of habitat destruction firsthand have turned into more active conservationists, but the train is about to leave the station. The urgency of the situation requires the help of everyone and the realization that all of us can make a difference. For example, there are little things that everyone can do, including buying only certified tropical wood and environmentally friendly coffee and cocoa, as well as recycling cell phones and other electronic devices that contain coltan, which is unfortunately found in abundance in areas inhabited by gorillas in the east of the Democratic Republic of the Congo.

At the same time, our daily life with the great apes in the forest shows us that there are solutions and that they can work. In Côte d'Ivoire, the many young people working with us are ambassadors in their villages. During the long tropical evenings, they share their knowledge and admiration for the chimpanzees, thereby helping to protect the habituated chimpanzees from poaching, as well as from the looting bands formed in the early 2000s, during the difficult years of civil unrest in the country. Through our commitment to a long-term project, we are at the same time becoming a long-term partner in the region. A real partnership can develop with the local population as well as with the national authorities in charge of national park protection. Anti-poaching campaigns become more effective, and more monkeys and forest antelope are often found closer to research stations than in other regions of a national park. In addition, all long-term research projects include local students, who later tend to become active advocates for conservation activities. Local researchers have become important forces guaranteeing the continuation of research projects whatever the political situation in the country, and are equally important as leaders spearheading conservation and research agendas in their countries. This also favors stronger collaboration between researchers and park management authorities, so that law enforcement can be directed to the regions within a park that most need it.

If we cannot save our fascinating cousins of the forest, what will our own future look like?

FURTHER READING

Boesch, Christophe. 2009. *The real chimpanzee: Sex strategies in the forest.* Cambridge: Cambridge University Press.

Goodall, Jane. 2010. *Through a window: My thirty years with the chimpanzees of Gombe.* Boston: Houghton Mifflin Harcourt, Mariner Books. 50th Anniversary of Gombe edition.

Nellemann, C., I. Redmond, and J. Refisch, eds. 2010. The last stand of the gorilla: Environmental crime and conflict in the Congo basin. A rapid response assessment. United Nations Environment Programme, GRID-Arendal.

Peterson, Dale, and Jane Goodall. 2000. *Visions of Caliban: On chimpanzees and people.* Athens, GA: University of Georgia Press.

Pusey, A.E., L. Pintea, M.L. Wilson, S. Kamenya, and J. Goodall. 2007. The contribution of long-term research at Gombe National Park to chimpanzee conservation. *Conservation Biology* 21:623–34.

Whiten, A., J. Goodall, W. McGrew, T. Nishida, V. Reynolds, Y. Sugiyama, C. Tutin, R. Wrangham, and C. Boesch. 1999. Cultures in chimpanzees. *Nature* 399:682–85.

Wrangham, R., and E. Ross. 2008. *Science and conservation in African forests: The benefits of long-term research.* Cambridge: Cambridge University Press.

CONTRIBUTORS

Christophe Boesch has been the director of the Department of Primatology at the Max Planck Institute for Evolutionary Anthropology since 1998. He is also the founder and president of the Wild Chimpanzee Foundation. He has studied a wide range of topics with focus on the Taï chimpanzees since 1979, including tool use, culture, hunting behavior, cognition, and social behavior.

Thomas Breuer works for the Wildlife Conservation Society as the principal technical adviser for the Nouabalé-Ndoki National Park in the Republic of the Congo. He has studied western gorillas at Mbeli Bai for the last eight years and holds a PhD in biology from the Max Planck Institute for Evolutionary Anthropology.

Chloé Cipolletta studied biology at the Sapienza University of Rome. After studying wild chimpanzees in Côte d'Ivoire, she worked for nine years in the Central African Republic, where she led the WWF Primate Habituation Programme, aiming to develop a controlled form of gorilla tourism in Dzanga-Sangha National Park. Presently she is the technical adviser for the WWF African Great Apes Programme based in Yaoundé, Cameroon.

Barbara Fruth is a research scientist of behavioral ecology at the Max Planck Institute for Evolutionary Anthropology. She has conducted research on wild bonobos in the Democratic Republic of the Congo since 1990 and codirects the Lui Kotale Bonobo Project. Since 2001 she has been research director of the Projet Cuvette Centrale, investigating floral diversity and medicinal plants with a focus on plant use by humans and apes.

Josephine Head is a PhD student in biology at the Max Planck Institute of Evolutionary Anthropology and project manager of the Loango Ape Project, Gabon. Her current research explores variation in dietary composition and habitat utilization between western gorillas and central chimpanzees, as well as the use of innovative, noninvasive methods for estimating ape densities.

Cleve Hicks conducted his doctoral work at the University of Amsterdam studying the Bili chimpanzees in the Democratic Republic of the Congo. He is currently at the Max Planck Institute for Evolutionary Anthropology and continues his research and conservation efforts in the Bili region of DRC.

Gottfried Hohmann is a senior researcher in the Department of Primatology at the Max Planck Institute for Evolutionary Anthropology. Together with Barbara Fruth he has conducted fieldwork in the Democratic Republic of the Congo for more than twenty years. Besides fieldwork in the DRC, his research experience includes field studies on macaques and langurs in Southeast Asia and experimental work on primate communication, social stress, nutrition, and digestive physiology.

David Morgan is a Research Fellow at the Fisher Center for the Study and Conservation of Apes at Lincoln Park Zoo in Chicago. His research explores the behavior and ecology of chimpanzees and gorillas and environmental factors shaping their shared coexistence in Central Africa. Most of his research takes place in the Goualougo Triangle, Republic of the Congo, a site he codirects with his wife, Crickette Sanz.

Martha M. Robbins is a senior researcher in the Department of Primatology at the Max Planck Institute for Evolutionary Anthropology. Her research interests focus on the impact of ecological conditions on social behavior, reproductive strategies, and population dynamics. In addition to her research on critically endangered mountain gorillas over the past twenty years, she has been involved with several projects studying western gorillas across Central Africa.

Crickette Sanz is an assistant professor in the Department of Anthropology at Washington University in Saint Louis. Her research focuses on chimpanzee material culture, ape sociality, and primate behavioral ecology. She and David Morgan are codirectors of the Goualougo Triangle Ape Project in the Nouabalé-Ndoki National Park, Republic of the Congo. The project's objectives involve long-term site-based research and conservation activities that address the major threats to great apes in the Congo Basin.

PLATE CREDITS

1, 11, and 12: Courtesy of Sonja Metzger

2, 5, 6, 7, 8, 9, and 10: © Caroline Deimel, Max Planck Institute for Evolutionary Anthropology

3: Cleve Hicks, The Wasmoeth Wildlife Foundation

4 and 13: Crickette Sanz, Goualougo Triangle Ape Project

14, 28, and 29: Ian Nichols, National Geographic Magazine

15, 16, 17, and 20: Courtesy of Thomas Breuer, Wildlife Conservation Society, Congo Program

18, 21, 22, 23, 24, 25, 26, and 31: Martha M. Robbins, Max Planck Institute for Evolutionary Anthropology

19 and 32: © Shelly Masi

27: Christophe Boesch, Max Planck Institute for Evolutionary Anthropology

30: © Barbara Fruth, Max Planck Institute for Evolutionary Anthropology

INDEX

Italic page numbers indicate illustrations.

Compositor:	BookMatters, Berkeley
Cartographer:	Bill Nelson
Text:	10/15 Janson Pro (Opentype)
Display:	Janson Pro (Opentype)
Indexer:	Alexander Trotter
Printer/Binder:	Thomson-Shore, Inc.

EATING APES
"The African Great Apes, our closest living relatives, are in imminent danger of extinction. Eating Apes, in beautiful prose, exposes the enormity and complexity of this conservation crisis. It took great courage to gather and present this information. You must read this book."
—JANE GOODALL
DALE PETERSON
KARL AMMANN

My Family Album
FRANS DE WAAL
THIRTY YEARS OF PRIMATE PHOTOGRAPHY

WORLD ATLAS OF
GREAT APES
AND THEIR CONSERVATION
JULIAN CALDECOTT
LERA MILES